⏺ PRAISE
MARKET FARMI
Revised and Expa

"We succeed at working this good
what we grow. No one offers better in
Byczynski. The marketing side of growing food needs attention as
much as soil prep. *Market Farming Success* doesn't miss a beat when
it comes to launching your hopes onto the local food scene."

—**Michael Phillips**, owner, Heartsong Farm,
and author of *The Holistic Orchard* and *The Apple Grower*

"With over twenty years of market-gardening experience and
teaching others the ins and outs of it via her wonderful publication
Growing for Market, Lynn Byczynski has created an up-to-date
guidebook for direct-market farmers. Whether you grow vegetables,
berries, herbs, plants, or other horticultural crops, *Market Farming
Success* is a practical must-read. I took home useful bits of advice
about sizing and building hoophouses, different trays and inserts
for seeding transplants, storing leftover seed, bed mulches, working
with restaurants, creating good intern relationships, types of insurance to consider as a market gardener—and found an appendix
full of places to seek out more information. The stories and photos
of farms around the country brought the information to life and
provided me with new ideas. This book will no doubt become a dog-
eared reference guide that is pulled from my bookshelf often as we
develop our new diversified farm."

—**Rebecca Thistlethwaite**, Sustain Consulting and author of
Farms with a Future: Creating and Growing a Sustainable Farm Business

"This overview of the business of market farming is a survival kit for new and aspiring vegetable farmers. It sheds light on esoteric Unknown Unknowns, and can save you from many pratfalls on the learning curve.

"Lynn explains each challenge of professional small-scale vegetable production in a calm, clear, confidence-boosting voice. She speaks from her own experience and includes sifted information gleaned from the many growers she knows as editor and publisher of *Growing for Market*.

"My own writing focuses on the planning and execution of crop production; the equally important marketing side is here covered by an extremely knowledgeable mentor."

—**Pam Dawling**, author of *Sustainable Market Farming*

"We've used the first edition of *Market Farming Success* in our beginning farmer program since 2007. This new edition contains additional information that makes it an even greater asset to those exploring and starting a market farm."

—**Adrian Card**, Colorado State University Extension

Market Farming Success

REVISED AND EXPANDED EDITION

Market Farming Success

✦ ✦ ✦

The Business of *Growing and Selling* Local Food

LYNN BYCZYNSKI

Chelsea Green Publishing
White River Junction, Vermont

Project Manager: Patricia Stone
Project Editor: Benjamin Watson
Copy Editor: Kim M. Smithgall
Proofreader: Alice Colwell
Indexer: Shana Milkie
Designer: Melissa Jacobson

Printed in the United States of America
First printing September, 2013
10 9 8 7 6 5 4 3 2 1   12 13 14 15 16

green
press
INITIATIVE

Chelsea Green Publishing is committed to preserving ancient forests and natural resources. We elected
to print this title on paper containing at least 10%
postconsumer recycled paper, processed chlorine-
free. As a result, for this printing, we have saved:

15 Trees (40' tall and 6-8" diameter)
7,195 Gallons of Wastewater
6 million BTUs Total Energy
481 Pounds of Solid Waste
1,327 Pounds of Greenhouse Gases

Chelsea Green Publishing made this paper choice
because we are a member of the Green Press
Initiative, a nonprofit program dedicated to supporting authors, publishers, and suppliers in their
efforts to reduce their use of fiber obtained from
endangered forests. For more information, visit
www.greenpressinitiative.org.

Environmental impact estimates were made using
the Environmental Defense Paper Calculator. For
more information visit: www.papercalculator.org.

Our Commitment to Green Publishing

Chelsea Green sees publishing as a tool for cultural change and ecological stewardship. We strive to align
our book manufacturing practices with our editorial mission and to reduce the impact of our business
enterprise on the environment. We print our books and catalogs on chlorine-free recycled paper, using
vegetable-based inks whenever possible. This book may cost slightly more because it was printed on paper
that contains recycled fiber, and we hope you'll agree that it's worth it. Chelsea Green is a member of the
Green Press Initiative (www.greenpressinitiative.org), a nonprofit coalition of publishers, manufacturers,
and authors working to protect the world's endangered forests and conserve natural resources. *Market
Farming Success* was printed on FSC®-certified paper supplied by RR Donnelley that contains at least 10
percent postconsumer recycled fiber.

Library of Congress Cataloging-in-Publication Data

Byczynski, Lynn, 1954–
   Market farming success : the business of growing and selling local food/ Lynn Byczynski.—Rev. and
expanded ed.
       p. cm.
   Other title: Business of growing and selling local food
   Includes bibliographical references and index.
   ISBN 978-1-60358-386-2 (pbk.)—ISBN 978-1-60358-493-7 (ebook)
1. Truck farming. 2. Vegetable gardening. 3. Farm management. 4. Farm management—Decision making.
I. Title. II. Title: Business ofgrowing and selling local food.

   SB321.B978 2013
   635—dc23

                        2013018632

Chelsea Green Publishing
85 North Main Street, Suite 120
White River Junction, VT 05001
(802) 295-6300
www.chelseagreen.com

MIX
Paper from
responsible sources
FSC® C101537

Contents

✖ ✖ ✖

Preface to the Second Edition

✖ ✖ ✖

If there's one certainty about market farming, it's that it changes all the time. In revising this book, first published in 2006, I have been impressed by the rapid pace of change. The single biggest difference since the first edition is that market farming has become trendy. Young people have arrived in droves, searching for meaningful work as farmers. Supportive organizations have proliferated to advocate for and assist small farmers. Market farmers have become media celebrities, featured in magazines and invited to write opinion pieces for major newspapers. Kids who grew up on organic farms go off to college and discover they now have a certain status among their peers. No longer do we have to explain what we mean when we say we're market farmers. The public now recognizes us.

Other trends have changed the face of market farming from within. Season extension with high tunnels and greenhouses has created much greater economic opportunity and allowed many part-time farmers to become full-time farmers. Winter markets have expanded accordingly. "Food hubs" have been developed in many places to help small farmers pool production to supply such big buyers as hospitals and college dining halls. Farm-to-school programs are gaining traction as public school systems buy local food for school lunch programs. Wholesale opportunities are abundant, and many small growers are scaling up to meet them.

Urban farming has taken off in the past seven years. Rural acreage is no longer a prerequisite for starting a farm; instead, many new growers start on abandoned city lots and suburban yards. Nonprofit farming has also become established, as food banks, watershed districts, sustainability organizations, schools and colleges, and many other groups have added food production to their missions.

New challenges have arrived on the scene, too. Food safety has become a much bigger issue for small farmers, with new regulations and certification demands from buyers. And of course there is the issue of climate change. I don't know many farmers who doubt that the weather is getting warmer and weirder. Frost dates, insects, disease pressure, irrigation requirements, and reliable varieties are less certain than in years past. A decade ago, "diversified" was considered the most important qualification for market farming success. Today that's still important, but there's an equally important characteristic: resilience. Farmers should be willing and able to change constantly, to keep pace with the constant changes around us.

—January 2013

Introduction

✖ ✖ ✖

Thomas Jefferson was a landowner, politician, diplomat, and president of the United States. But in his heart, he claimed, he really wanted to be a market gardener.

"I have often thought that if heaven had given me choice of my position and calling," Jefferson wrote, "it should have been on a rich spot of earth, well watered, and near a good market for the productions of the garden. No occupation is so delightful to me as the culture of the earth, and no culture comparable to that of the garden."

I have always loved that quote, because Jefferson expressed so well the passion that I and many others have felt about gardening, a passion that can be satisfied only by a commercial-scale garden. At some point in your progress as a gardener, you probably have found yourself with far too many tomatoes or zucchinis or bedding plants to suit the needs of a single family. But it's as though you can't stop yourself from growing more and more every year, until you realize you could be selling your produce.

Welcome to the club. I felt the first horticultural longings back in college, but suppressed them in favor of a career as a newspaper reporter. But my thoughts kept returning to the idea of growing herbs, vegetables, or flowers as a business, and I avidly read every book and magazine article I could find on market gardening. Then I met a man who had just bought a small farm, complete with a gingerbread farmhouse, big dairy barn, a pond, and 20 acres. My soulmate! We married, and within a year started growing vegetables to sell at the farmers market. The first few years were hard; we were thwarted by drought and grasshopper populations that I have never seen since. But we were full of optimism and energy, and we loved—truly loved—the hard work and purposefulness of market gardening.

My biggest complaint in those years was a feeling of isolation. We didn't know any other organic market gardeners, and we didn't feel that anyone else could really relate to the life we were living.

Four Season Farm in Harborside, Maine, is a model for many of the small market farms in the United States. Owners Eliot Coleman and Barbara Damrosch are renowned for dedication to their craft and many important innovations, such as movable high tunnels. Coleman's books *The New Organic Grower* and *The Winter Harvest Handbook* are essential reading for all beginning growers.

We didn't have an organic farming association in our state back in the 1980s, and there was little opportunity to meet and learn from other growers. To fill that hole, I decided to use my journalism skills to start a national publication for market gardeners, and *Growing for Market* was born in January 1992.

In the ensuing years, growing for market and *Growing for Market* have been my twin occupations, each feeding the other and keeping us going body and soul. In the beginning, the magazine was just a sideline to our farming business, but as it became more successful, it took up more of my time and, consequently, farming took up less. My husband and I have continued to farm at varying levels of intensity while working other jobs. We have sold at farmers markets, been founding members of the country's first cooperative community-supported agriculture (CSA) group, sold to chefs and the local food co-op, grown flowers for florists, and on and on. As I write this at the beginning of 2013, we are taking a hiatus from farming, and I have to confess I'm a little worried about exactly how I am going to react when the days get longer, the trees start to bud, and another growing season beckons. Once farming gets in your blood, it's not easily ignored.

I have talked to countless other people who have been bitten by the market farming bug. I know of one Academy Award–winning actress, one major rock star, and one software billionaire who have started market farms. I know builders, college professors, computer experts, doctors, corporate executives, TV meteorologists, and others who achieved great status in their professions but gave it up to farm. I know big farmers who scaled down to market farming—and became more profitable. I know young people who went to college for advanced degrees, then decided they would rather work on a farm. Market farmers run the gamut from teenagers to people in their 80s. They come from all socioeconomic backgrounds and from every state in the nation, including many places that you don't think of as being hospitable to vegetable and flower production. *Growing for Market* has subscribers in every part of North America, from the US Virgin Islands to the Yukon; from New York City to the coast of California and every state in between. Whatever your background, wherever you are located, rest assured that you can be a market farmer.

What to Call Yourself

You'll find there is some variation in the words commercial gardeners use to describe their occupation: market gardener, market farmer, direct-market farmer, vegetable farmer, and truck farmer are all part of the lexicon. There are no official definitions, but here are distinctions made by John Hendrickson of the University of Wisconsin:

- **Market gardens** have fewer than 3 acres in production, not counting fallow or cover-cropped areas. Market gardeners use mostly hand labor.
- **Market farms** have between 3 and 12 acres in production, not counting fallow or cover-cropped areas. Market farmers use a mix of hand labor and mechanization.
- **Vegetable farms** produce crops on more than 12 acres, which requires mechanization.

The phrase "market garden" was in widespread use in the late 1800s, when seedsman Peter Henderson wrote the classic book *Gardening for Profit*. (His book is well worth reading today, certainly for learning about your heritage as a market gardener, but also because nothing ever really changes; much of the advice he gave in 1886 is still applicable.) Henderson refers to himself as a market gardener, even though he grew more than 10 acres of vegetables. To him, the term referred to someone who grew a wide variety of produce to sell.

The term "truck farmer" is not used much anymore, but many older people will recognize it as referring to people who farmed on the outskirts of cities and trucked their produce into the city to sell. "Direct-market farmer" is an all-purpose phrase that describes anyone who grows something that is sold directly to customers, rather than into a wholesale or processing chain. Baby boomers who grew up on farms, even large commodity crop farms, remember some direct-marketing activities taking place. Their mothers may have sold eggs or milk from the family's hens and milk cow, or the children may have sold garden extras at a roadside stand. That is uncommon now on large crop farms because rural populations have declined as farm size has increased, meaning there are not many neighbors to buy products directly from a farmer. But many large vegetable farms in recent years have created a direct-marketing sideline by selling at farmers markets in nearby cities.

The phrases "agritourism," "destination farm," and "entertainment farm" have cropped up recently to describe operations that attract customers to the farm with retail stores, pumpkin patches, hay rides, and more elaborate attractions, such as corn mazes, festivals, and pizza gardens.

What Does It Take?

Whatever their age or background, and whatever they call themselves, people who get into market farming have many things in common. Everyone quickly finds out that market farming is one of the most complicated and challenging jobs you could ever hold. You have to be a good grower, which is no simple matter itself, given all the considerations of soil, scheduling, variety selection, crop management, and harvest. And you have to be able to handle all those tasks not for one or two crops, but for literally dozens of crops.

Second, you have to be good at marketing—to know how to advertise your produce, price it, display it, educate people about using it, and cross-market with other products you're selling. Finally, you need to have a head for business—to know how to keep records, pay your taxes, know your production costs, stay informed about relevant laws and regulations, buy insurance, hire help, and much, much more.

This book isn't going to tell you how to do all those things, because, if it did, it would be as big as an encyclopedia. There are already many fine books and free publications that tell you how to do each and every one of the jobs required of a market gardener. Instead of attempting to cover all aspects of market gardening, this book will do two things:

1. Identify the topics you need to know to get started; in particular, it will explain the issues that make the difference between success and struggle on a new farm.
2. Identify the resources that already exist on each of these topics and tell you how to find them.

What Will You Find in This Book?

As a longtime market gardener myself and the editor of *Growing for Market* magazine since I founded it in 1992, I have learned that most beginners don't lack knowledge about growing; what they do lack is an understanding of how to grow on a **commercial** scale. They don't understand the planning, budgeting, marketing, and management aspects of market gardening. They don't know the inside secrets of the business, such as where to get the best deals on supplies and equipment. And they don't know what information is already out there to tell them.

I wrote this book in an attempt to cut through some of the mysteries of commercial horticulture. When we started market gardening in 1987, we didn't even know the names of greenhouse companies or packaging suppliers—and we didn't know how to find out. Today, of course, the Internet makes it possible to find virtually anything, and should be your first stop as you search for information. But the Internet has the disadvantage of giving you *too much* information, especially if you don't know the exact name of the product you need, or the commonly used term for a problem you're having. If you put "greenhouse" into a search engine, for example, you are going to pull up 77 million web pages, which is not a great help.

This book will introduce you to aspects of commercial growing that differ from backyard gardening. It will give you the language of market farming, and explain the terms that you are expected to know. Armed with the basic concepts and terminology, you can turn to the wealth of free information that is on the web. For example, once you know what's in a "crop enterprise budget," you can Google that phrase with the name of your state or the crop you want to grow. Or you can go to YouTube to look for videos about "stale seedbed" or "caterpillar tunnel" or just about any other production practice you'll learn about in these pages.

My hope is that you will be a more sophisticated market gardener after reading this book, that the information here will help you move quickly and confidently through the inevitable learning curve, and give you a quick start on the path to success.

CHAPTER 1

Getting Started in Market Farming

If you're reading this book, chances are good that you are considering a market gardening business for one or more of the following reasons: you love to grow plants, appreciate good food, want to be your own boss, enjoy working outdoors, or want to have a direct hand in changing the world through a more sustainable agriculture.

Those are the most common reasons market gardeners express about why they are involved in this business. No one ever says they're doing it for the money. The simple fact is that farming is not a high-dollar occupation. It hasn't been for at least the past century, nor is it today. Nevertheless, farming continues to beckon people in every generation, offering the promise of a better life—a life that is more peaceful and productive. The benefits of farming are enormous, but most of them don't show up on financial reports. That's not to say that the benefits of farm life don't have a price tag attached. Certainly, there are financial advantages to having good physical and mental health, for example. And there are many tax savings associated with working your land as a farm. But the primary benefit of farming is not wealth.

How Much Money Can You Make?

Even if you accept the fact that farming is not a high-paying occupation, even if money is not highly important to you, you still have to think about it when you're starting out. You need to know how much it's going to cost to get started. You need to know how much you can potentially earn once your farm is established. You especially need to know whether to hang on to another job in the meantime. As you develop your new farming business, you should make financial viability one of the tenets of your planning. A business can be sustainable only if it makes enough money to meet your financial needs.

Financial needs differ from one farmer to the next. A family of five has different financial needs than those of a couple without children. Someone who wants to make his or her entire livelihood on the farm will have a different perspective on profits than someone who views farming as a sideline. I'm not going to tell you how much you need to make to be considered a success. That's entirely up to you. Over the years, my definition of a "sustainable farm" has broadened, and I now think that the person who keeps farming obviously has achieved a satisfactory measure of financial sustainability.

In my role as editor and publisher of *Growing for Market* (*GFM*) magazine, I have discussed finances with a large number of market gardeners. Some have been willing to share financial statements with *GFM* readers, and others have told me their income but didn't want it published. As a result of these discussions, I have developed a clear picture of the earning potential of market farms.

Farmers who are successful—that is, making enough money to keep farming—can have an operation of any size, from a tiny, part-time start-up to a large, established business. They are growers who have achieved a balance between income and expenses, carving out enough to pay themselves fairly while building equity in land, buildings, and equipment.

At one end of the scale are growers who pay themselves the same wages as their employees, sometimes as little as minimum wage. At the other end of the scale are people who net $100,000 or more per year—but often that represents the work of both spouses, so the per-person income in even the high-end situations is modest, though certainly adequate.

However, market gardening clearly offers something that money can't buy, because none of the veteran market gardeners I have interviewed expressed any interest in quitting to take a more lucrative job. For most farmers, the financial goal is to make enough money to live on and put a little away for retirement, while doing work they love, spending time with family, and making a contribution to the community.

In the sections below, you will read more details about finances on several different types of farms. The variables from one farm to another are numerous, so it's not possible to state unequivocally

Should You Quit Your Day Job?

If you ask a dozen veteran growers this question, you're likely to get six who say, "Yes!" and six who say, "No!" In other words, people are divided over the correct advice to give a beginning farmer.

Reasons you should quit: There's the sink-or-swim theory, which holds that you will be forced to succeed if the alternative is bankruptcy, poverty, or getting another job. The other chief argument is that if you are going to be serious about farming, you need to do it full-time, every day of the week, to produce the kind of quality that will make you successful.

Reasons not to quit: With an outside income, you can invest in your farm over time until you have the tools, equipment, and facilities you need. You will not be saddling yourself with debt that will keep your farm unprofitable for much longer than necessary. You will feel more secure and make better decisions. Your family may have health insurance.

My opinion falls somewhere between the two. The ideal situation is to spend a few years working on other farms and learning as much as you possibly can about every aspect of market farming. But you especially need to learn how to grow. There is no substitute for experience, particularly with a good

farmer as a teacher. You can teach yourself marketing and business management, but you need to actually grow vegetables and flowers for several years before you will learn those intricate skills. At the very least, you should grow the biggest garden possible and keep records about everything you do.

When you decide to start your own farm, you would be smart to have some source of outside income—either a spouse with a job (and health insurance) or a freelance/seasonal job. The vast majority of American farmers have some source of outside income, and market farmers are no exception. You might do remodeling or house painting in the slow season; drive a school bus twice a day; substitute teach; do web and graphic design; consult for former employers from home; or any similarly flexible work. Look for revenue streams, both within your farming business and outside of it, that will combine to keep your boat afloat when getting started.

When you know you want to be a full-time farmer, and feel you have the necessary skills, write a business plan and show it to a local Small Business Development Center advisor. If you feel reasonably assured that you can succeed as a farmer, quit your city job and go for it!

that if you follow one model you will make a certain amount of money. Length of growing season, proximity to markets, growing expertise, marketing skill, weather disasters, and many other factors influence revenue on individual farms. The dollar amounts in the examples listed below represent what is possible on farms that are well managed by experienced growers, in hospitable growing conditions. I will group them by the previously mentioned categories—fewer than 3 acres, 3 to 12 acres, and more than 12 acres.

These are just approximate sizes for purposes of discussion; there are always going to be exceptions to every category.

For the purpose of applying other farmers' numbers to your own farm, the two most important figures are the gross revenue per acre and the margin, which is the percentage of revenue that is left after expenses. The gross per acre multiplied by the number of acres farmed provides gross revenue; that, multiplied by the margin, provides the net income. On a family-owned farm, net income is the same thing as the farmer's pre-tax salary.

▶ FULL-TIME FARMING ◀

If you want farming to be a full-time livelihood, you need to have realistic expectations about how much money you can make. Start by calculating how much you *need* to make to provide for yourself and your family. The descriptions below will help you determine what scale your farm should be to meet your income goals.

Fewer than 3 acres

A rule of thumb in market gardening is that one person working full-time can handle about 1 acre of intensive production. In this model, at least one person, and often two, work full-time on the farm with little or no hired help. They grow a wide array of crops, but with a particular focus on high-dollar crops, such as salad mix, heirloom tomatoes, and cut flowers; and they sell in diverse markets, including farmers markets, to restaurants, and through a community-supported agriculture (CSA) component.

The amount of money that can be earned per acre on this type of farm varies considerably, based on the length of the growing season and differences in management practices. It could be $20,000 per acre for mixed vegetables to $35,000 an acre or more for high-dollar salad mix, herbs, or cut flowers. Whatever the per-acre revenue, the margin on this type of farm consistently runs at about 50 to 60 percent, which is considered a very good margin. At this scale, farmers rarely hire labor, preferring to do the work themselves rather than managing other people. They purchase only basic tools and equipment. The bulk of expenses on this type of farm are for seeds, plants, and supplies.

3 to 12 acres

Many growers who started with just a few acres soon find that they need to grow more to earn enough for a full-time livelihood. It's

Harmony Valley Farm in Viroqua, Wisconsin, is one of the most admired organic farms in the United States. Started in 1985 by Richard de Wilde in a secluded, spring-fed valley in the western part of the state, the farm produces more than 100 acres of certified-organic produce, grass-fed Angus beef, and pastured pork. Harmony Valley Farm sells at the Dane County Farmers Market in Madison, through a CSA, and to retailers and restaurants across the Upper Midwest. PHOTOGRAPH COURTESY OF RICHARD DE WILDE, HARMONY VALLEY FARM

impractical to grow more than 3 acres of produce using just hand labor, so farmers invest in labor-saving equipment, which means higher capital costs, depreciation, maintenance, and repairs. Hired help will also be necessary.

The amount of revenue generated by this size farm depends tremendously on the grower's energy and marketing abilities. To sell more than 5 or 6 acres of vegetables at retail prices is a feat requiring attendance at numerous farmers markets each week, or an on-farm market. For most growers at this scale, wholesaling to grocery stores, restaurants, and institutions is part of the marketing mix.

More than 12 acres

At this scale, mechanization is essential, and the gross per acre is much lower than on small farms, simply because production isn't as intensive. Plants need more space to allow tractors, transplanters, cultivators, and harvesters to get through, so the number of plants per acre is smaller than on a hand-tended vegetable field. In addition, the greater the production on large acreage, the less

A Full-Time Livelihood

Chip and Susan Planck of Wheatland Vegetable Farms in Purcellville, Virginia, have always have been willing to share information about their farm's finances, in the belief that it helps other growers. Both worked full-time as vegetable farmers from 1973 until 2010. For the majority of those years, they grew 20 acres of vegetables, using ecological methods only, and sold their produce exclusively at 12 producer-only farmers markets in the Washington, D.C., metro area. Annual gross sales were $300,000 per year. Labor expenses were $100,000, for about two dozen young people per year, ranging from four to 16 people at any one point in their seven-month season. They paid the hourly federal minimum wage, plus provided on-farm housing, utilities, and vegetables. All nonlabor expenses combined totaled another $100,000, leaving $100,000 net income for Chip and Susan.

Although spending a third of expenses on labor may seem like a lot, Chip and Susan feel that it is the key to their financial success.

"This is labor-intensive agriculture for three, possibly other, reasons," Chip said. "One, the crops that people want are not susceptible to machine harvest (heirlooms and low-fiber beans, for example); two, we direct-market to avoid the 'middlemen' and hence need people for that packing, hauling, and selling; and three, insofar as sustainable methods are part of the customer appeal, some of them are more labor-intensive."

From 2007 to 2010, the Plancks scaled back their production to 5 and then 2 acres per season. They found that their gross income declined—to $200,000 and $80,000, respectively—but the gross dollars per acre doubled. During this period, Chip and Susan rented portions of their farm, with equipment and infrastructure including deer fencing, greenhouses, irrigation, cooler, and worker housing, to farmers wishing to run their own businesses. Their farmers market managers were willing to allow the independent farmers to sell in the Plancks' vacated space.

For those renting for three years, the annual charge was $1,500 per acre and 3 percent of gross. Rental income to the Plancks was about $15,000 per farmer per year. Average rent cost to farmers was 9.5 percent of gross. Two of the four renters

likelihood the farmer can sell it all direct to the consumer at retail prices. Revenue per acre may be as low as $10,000, but with 12 or more acres in production, the gross revenue is high.

But is the net revenue any higher on these farms than on the small, intensive farms? Often, it is. On most of the larger farms I have visited, profit margins range from 10 percent to nearly 50 percent. In addition, most of these larger farms are more meticulous in their record-keeping than smaller farmers and have already deducted depreciation and taxes as an expense before they cite their stayed for three years each, accumulating business records allowing them to qualify for USDA/FSA beginning farmer loans for mortgages on their own farms.

Chip and Susan subdivided their 60-acre farm using a county ordinance for cluster zoning. This allowed them to permanently ease from development 58 acres while creating a hamlet of seven lots, ¼ to ⅓ acre each, on 2 acres. These lots are surrounded by 8 acres of common open space. The remaining 50 acres have been sold to like-minded farmers.

The Plancks are now turning their attention to the sale of the hamlet lots. In addition, they are working with a regional land preservation organization and with several other knowledgeable vegetable and animal farmers to purchase and ease from development larger acreage on which to outfit an incubator farm where ready-to-start farmers could rent portions of the land and infrastructure to run their own operations.

The Plancks' influence in the market farming world has been extensive. Twenty-five of their 250 former workers have gone on to run their own farms. Their daughter, Nina

Planck, created London Farmers' Markets—a dozen producer-only markets in England—and a farmers market in Washington, D.C. She is the author of the book *Real Food: What to Eat and Why*, which documents that whole foods, including meat and dairy, that are grown, raised, and prepared with traditional methods, are essential for human health.

net income, so they have a more accurate figure for their personal earnings. More on those considerations later.

❧ THE SIDELINE FARM ❧

Many market gardens are operated as part-time enterprises, with the farmers holding other jobs either on or off the farm. Because growing produce and flowers requires a high level of attention, it's difficult for the part-time farmer to be as efficient and productive

A Part-Time Farm

My own farm provides a good illustration of the sideline model, and I know from conversations with many other part-time growers that our revenues are on a par with others in our situation. Here are two examples of our farm's income:

* In one of our busiest years, we grew vegetables and cut flowers on about 4 acres and hired three people to help us part-time during the peak summer months. We sold to florists, restaurants, natural foods stores, and through a cooperative CSA. Total sales were about $46,000. Expenses were $28,500, including about $8,000 in wages and payroll taxes. That's a net of about 37 percent of income—or $17,000.

* As we started to scale back, we grew only cut flowers on about 1 acre, and sold them to florists and a natural foods store. Total sales were $32,000. Expenses were about $13,000, including $2,250 in wages and payroll taxes for our one employee who worked part-time for two months. Our net income was 59 percent of gross, or $19,000.

In both cases, the net income was a nice supplement to our off-farm income, and the equivalent of about $15 per hour for our work, most of which was enjoyable and personally fulfilling. Furthermore, farming our land does provide other financial benefits, such as lower property taxes and federal income tax deductions on many other possessions. You'll read more about that in chapter 8.

as the farmer who is always present on the farm. As a result, gross revenue per acre will be somewhat less than that earned by the full-time farmer. And expenses will probably be higher, because the part-time farmer will have to hire help to get the work done.

Despite those constraints, it is not unusual for a part-time farmer to gross $10,000 to $15,000 an acre on produce and flowers and to net about half that amount. Net revenue on the start-up farm is a nebulous figure, however, because many growers just plow their profits right back into the farm. For the first few years, and maybe for much longer, the part-time grower realizes that he or she isn't big enough, and so reinvests the farm income in greenhouses, tractors, tools, land, marketing aids, and experimental crops.

And that's as it should be. Debt can sink a fledgling farmer who has a bad year, so it's advisable to grow the business slowly until all the pieces are in place. Once a grower has experience and skills, it may be time to take the leap into borrowing money to scale up the business.

Finding Your Farm

If you could farm anywhere in the country, where would it be? For most people, the answer is "right here." Family, friends, familiar landscapes, and other emotional factors determine where we settle. However, I do know a few farmers who have chosen their locations based on the likelihood of success. They have looked for areas of the country with a good growing climate, a developing population base, a community that embraces small farms and locally grown food, and reasonably priced land. Alex and Betsy Hitt of Peregrine Farm in Graham, North Carolina, graduated from college in Utah and went searching for the best place to be market farmers. They eventually decided to head to the Research Triangle area of North Carolina, and bought a small farm about 40 minutes from that burgeoning urban area. Their careful research paid off—they have been successful, full-time market farmers at the same location for more than 30 years. Other people have headed for areas that are well known as great places to grow horticultural crops, such as the California coast and the Willamette Valley of Oregon, and have met with similar success.

But every location has its advantages and disadvantages, even those dream locales, and on balance you may find that you can do just as well farming in the place that draws you for reasons besides farming—the place where your family owns land, the town where your parents and siblings live, or the place where you went to college.

If you haven't yet bought or leased land, so much the better. You can choose a farm that will help make your farming business a financial success by considering the factors listed in this chapter. If you are already settled on a piece of land, you may have to make some changes to optimize the farm's prospects for success. In either case, here are some things to keep in mind about the place where you farm.

► THE IDEAL FARM ◄

If I could create the perfect small market farm, it wouldn't look much different from the farm where I live now, so let me tell you what's good about my farm—and what would make it better.

First, it's a manageable scale. We own 20 acres, about half of it cropland, the other half covered by our home and outbuildings, two

The author's farm is laid out with grass paths between the beds for easier planting and harvesting. With only 2 to 4 acres in production, there's plenty of room to spread out on the 20-acre farm.

ponds, and a hillside meadow. For small-scale market farming, 10 production acres are ample for cash crops and soil-building rotations. The land is south-facing, so it warms up early in spring, with a wooded hillside to the north that protects us from winter winds. It has good soil, very high in organic matter and natural fertility. Our soil tests rarely recommend significant amendments for plant nutrition. Our farm ground has a 5 percent slope, so it drains well in rainy seasons. We are in an agricultural valley, so we still have a great bucolic view to enjoy every day.

Our farm is on a paved road, which saves a lot on flat tires and makes it easy for customers to reach us. We are 7 miles from the city limits of a growing university town of 100,000 people, and 40 miles in either direction from larger cities, which gives us easy access and receptive markets for our produce.

Our farm also has a nice farmhouse, a big old dairy barn that we use for machinery and equipment storage, and a smaller barn that we use as a packing shed. Although the entire place is sloped in two directions, we have enough almost-flat land that we've been able to put up five greenhouses.

The single biggest shortcoming of our land is the lack of windbreaks around our fields. We live in a windy state, and we have always considered the wind our biggest obstacle to production. Being surrounded by row crops, too, could create a problem with drift from our neighboring farmers' chemical sprays, but they have always taken care to spray only when it's calm, except for one instance when we had to ask the neighbor to stop spraying herbicide. (He got mad, but he did stop.) If we could change anything about our farm, we would have tall windbreaks on every side of the fields, but especially on the south side, where the prevailing wind originates.

Soil quality

Your first consideration when buying land should be the soil. Soil quality can be the single biggest factor in your success as a farmer, so do your research before you go out to look at a piece of land.

First, familiarize yourself with the characteristic soils in your target area, specifically those recommended for vegetable production. If you haven't already, go to the local Natural Resources Conservation Service (NRCS) office or county Extension office and get a copy of the soil survey for the county where you are thinking of buying or leasing land. Soil survey maps in amazing detail are available for most of the United States. This information is also available on the Internet in an easy-to-search format. Go to websoilsurvey .nrcs.usda.gov.

Every type of soil has a name and a description, including information about its suitability for specific agricultural purposes. Learn the names of the soils that are suitable for vegetable production (your Extension or NRCS agent can help you if you're unsure), and find those soils on the map. Or find the area you are most interested in, and check the key to see what the soils are like there. This will give you a broad picture of where to find the best soils. It will also help you narrow your search so that you don't have to run off to look at every farm or parcel of land that comes on the market. That way, you won't be swayed by a pretty farmhouse, winding lane, splashing steam, or other attributes that might not help at all in market farming.

Shop first for good soil—or at least soil that can be improved. Few market gardeners start with perfect soil. Most have to work hard to build soil fertility and tilth; some have to bulldoze trees to create fields. The issues that really can't be resolved are the physical ones, including drainage, texture, and slope. You need adequate amounts of soil—don't assume you can truck in soil, because you'll soon find out that it's too expensive and a never-ending proposition. Avoid land with rocky ground for horticultural crops. You'll wear out yourself and your equipment dealing with rocks. Chemical contamination can't be remedied, so find out what the land was used for in the past. Talk with established growers in the area. What

The Best Soils

You will often hear people talk about "prime farmland," which is a term that is descriptive but still somewhat vague. It's defined this way: *Prime farmland* is land that has the best combination of physical and chemical characteristics for producing food, feed, forage, fiber, and oilseed crops and that is available for these uses. It has the combination of soil properties, growing season, and moisture supply needed to produce sustained high yields of crops in an economic manner if it is treated and managed according to acceptable farming methods. In general, prime farmland has an adequate and dependable water supply from precipitation or irrigation, a favorable temperature and growing season, an acceptable level of acidity or alkalinity, an acceptable content of salt or sodium, and few or no rocks. Its soils are permeable to water and air. Prime farmland is not excessively eroded or saturated with water for long periods of time, and it either does not flood frequently during the growing season or is protected from flooding.

Many types of soils fall into the category of prime farmland. Perhaps a more useful terminology would be the "capability" classes assigned by the Natural Resources Conservation Service (NRCS). Class 1 soils are the best—they have very few limitations that restrict their use for growing crops. Class 2, 3, and 4 soils can be used for market farming, but you need to be aware of their limitations. Anything higher than Class 4 is suited only for pasture, not for crops. Here are the official NRCS descriptions of those first four soil classes:

Class 1 soils have slight limitations that restrict their use.

Class 2 soils have moderate limitations that reduce the choice of plants or require moderate conservation practices.

Class 3 soils have severe limitations that reduce the choice of plants or require special conservation practices, or both.

Class 4 soils have very severe limitations that restrict the choice of plants or require very careful management, or both.

do they consider good soils? Where would they buy land? Do they know of already established or cultivated places for sale?

When you have found land that seems to have good potential, take a shovel when you visit and dig an 18-inch-deep hole every place that you think might make a good crop field. Look at the depth of the topsoil, and look for impervious layers of clay. Feel the texture of the soil, roll it into a ball, get it wet. If you're unsure of your own judgment, take a sample to an Extension office for some expert advice on the structure. While you're there, pick up a brochure about how to take a sample for a soil test—you will need to do one every year or two once you start farming.

The Web Soil Survey is a quick way to identify the type of soils in a given area. The orange lines show broad areas of each soil type, which are named in the panel at left. Detailed information about each soil type is also available on the survey.

Other features

No farm is perfect, but some are much nicer than others. And sometimes the things you thought were benefits can turn out to be drawbacks, and vice versa. Here are some of my thoughts about what makes a piece of land a good candidate for a small-scale, sustainable farm.

* After soil, water is the second most critical feature. You must have access to clean, ample water in order to farm. Food safety programs, which will be explained in great detail later, require that you use potable water to irrigate and wash food crops. The most straightforward water source is a municipal or rural water system. If water lines exist or can be installed for a reasonable price, consider this option first, because the water supplier does all the work for you—testing water quality, regulating water pressure, even identifying the source of leaks that might occur on the farm. Many people assume it will be much too expensive, but that may not be true. Find out the price of water before you make a decision, and keep in mind that even if you spend a few hundred dollars a month on a water bill, it will be for only a few months of the year, and the value of the crops

you'll be growing will justify the expense. For the small-scale, intensively worked farm, using municipal water ultimately may be the cheapest solution. As the farm gets larger, that may change. Ponds, streams, and wells are also possible sources of irrigation water. Get it tested to be sure it's safe to use for irrigating vegetable crops. Be sure there is enough water and enough water pressure to irrigate. The typical water pressure entering a home is 40 to 80 psi (pounds per square inch). Drip irrigation, which is preferred for most vegetable and fruit crops, requires 15 to 30 psi. If there is a stream or river on the property, be sure you have the right to use the water—in some cases, water rights are allocated, and adjoining landowners are not necessarily entitled to use the water.

❋ As a general guideline, be sure you will be able to irrigate crops before you buy or lease land; you cannot depend on rainfall to keep your crops thriving, so a reliable water source is essential. Your state Extension service should be able to help with irrigation planning, whatever the water source.

A reliable source of clean water is essential for irrigating a market garden. Most growers use municipal water or groundwater. A pond or stream can work if it's not downhill from a source of contamination, such as livestock. Even then, irrigation water will have to be tested to comply with food safety standards. Surface water should not be used for washing produce.

❀ Farming history can be important, especially if the land has been farmed conventionally with herbicides, pesticides, and other chemicals. If you want to farm organically, you're going to have to wait three years to get your land certified if it was previously treated with chemicals. And some chemicals persist in the environment for a long time—not a reassuring thought to someone who is looking for a healthy job.

Some previously used chemicals can harm your crops; for example, the herbicide clopyralid has been used to kill weeds in agricultural crops including wheat, timothy hay, corn, sugar beets, and others. It is widely used by lawn-care companies. Clopyralid and other herbicides, including picloram and aminopyralid, do not biodegrade quickly, even when composted. Small amounts of the residual chemical can damage many horticultural crops such as tomatoes, beans, peas, and potatoes. Ever since municipal compost in Washington State was found to be widely contaminated by clopyralid, horticulturists have been advising small farmers to be careful about the source of any compost or manure used on the land. Similarly, if you are buying land that has been farmed conventionally, you should ask about previous herbicide use. Unfortunately, clopyralid is the active ingredient in more than 40 brand-name products. For more information on clopyralid, see chapter 6.

Another potential problem on farmland is excessive salts in the soil from livestock. The problem is easily revealed by a soil test, which you should have performed as soon as you find land you're serious about. The county Extension office will perform a simple nutrient test for a few dollars, though you may have to specially request and pay extra for the soil to be tested for soluble salts and organic matter. More complete tests can be conducted later to tell you what amendments are required for the types of crops you are planning to grow.

❀ Natural features of the land can greatly affect the success of a farming venture. You need open land that is well drained, but not so hilly that soil erosion will be a problem. A south-facing slope is ideal for vegetables and flowers because it warms up earlier in spring. The fields should have a windbreak from the prevailing wind, but being surrounded by forest often suggests you'll have problems with deer and may need to erect a deer fence.

You also should check to be sure that the lay of the land doesn't pose potential problems with runoff from adjacent farms, especially livestock farms. You may need to get your farm certified under food safety regulations in the future, which require that produce is not contaminated by manure, septic leach, or other kinds of contaminants running off nearby fields.

❀ Neighbors can be a blessing or a problem for a small farm. Find out all you can about anyone within hearing distance, to determine if your potential farming activities might disturb them. It's hard to believe that something as natural and wholesome as market farming could be objectionable, but more than one grower has had run-ins with neighbors who complained about livestock, tractors, greenhouse fans, or employees.

❀ Location in relation to markets is an important consideration. Highway frontage can be a benefit if you want to have a farm store or retail greenhouse; otherwise, traffic noise, litter, and pollution may detract from your pleasure in working outdoors. Being way off the beaten path, though, has drawbacks, too, such as difficulty getting supplies delivered, excessive time spent driving to town, and wear and tear on your vehicles. Being too far from your markets is a disadvantage, especially if you are going to be selling at a farmers market that starts early in the morning.

➤ WHERE TO LOOK FOR LAND ◄

Finding farmland is not as easy as finding a house, and there are several important resources beyond the obvious route of contacting a real estate agent.

Your first step should be to devote some research time to Internet real estate listings. Search for "farms for sale" plus your desired location. The search engine will return some national sites, such as landandfarm.com, as well as local real estate agencies. Although you are not likely to find the farm of your dreams this way, you will learn a lot of information about prices and features that are considered desirable. For example, when we went looking for our current farm, we immediately noticed how many listings mentioned "Lawrence Schools" as a selling point. That was a clue to do some research on public schools in the area, since we had young children. You'll also begin to understand which agencies and agents specialize in the kind of land you're seeking. Sign up for updates with the person who seems most on top of the farm situation in your area. Explain what you're looking for carefully, as some people may not understand the terms "market garden" or "market farm." You'll probably find that the phrase "hobby farm" elicits recognition from some real estate agents. (Apart from this one occasion, never call your farm a "hobby farm" again, as you will learn in chapter 8.)

Another place to look for a farm is in the agricultural publications that serve the area. Statewide organic and sustainable agriculture associations publish magazines or newsletters, and the classified ads in these publications are a goldmine of farms for sale. You may even want to place a Farm Wanted ad in this kind of publication.

Once you've searched in the obvious places, you may find that it pays to do your own legwork. Land transactions often occur among neighbors, without the involvement of a real estate agent. We got our current farm by driving around and looking at places we thought would be good for a vegetable and flower farm. Then we looked up the owners' names and addresses at the county assessor's office and wrote letters to those landowners telling them we were looking for a small farm and would appreciate hearing from them if they were thinking of selling their land or knew of other possible farms that might be coming onto the market soon. One farm in particular we identified as our "dream farm" because of its location, south-facing slope, house, and barns. A month after we sent our letter, the owners of our dream farm called to say they were going to sell it. It turned out that their dream farm had just become available when a woman 2 miles up the road heard that her dream farm was for sale. When we made our moving plans known, a friend immediately offered to buy our old farm. Four families bought new farms, without one of us advertising or paying a single dollar of real estate commissions.

Our experience may have been an anomaly, but there are plenty of reasons that farmland might be available even if it's not listed for sale. Many times, people aren't really thinking about selling until someone shows an interest in their farm—and then, suddenly, they start thinking that maybe it is time to move into town or sell off a piece of their land. Other farmers are dedicated to preserving their land for farming, and they would never consider selling it to a developer or even someone who just wanted a big house in the country. But they might feel quite differently about selling to someone who wants to farm it.

So explore all channels for finding a good piece of farmland. Contact real estate agents, talk to the county Extension agent, introduce yourself to landowners by mail, tell other farmers you meet that you are looking for land, and just get your name out on the grapevine. In most rural areas, community ties are still strong and word travels fast. With luck, the perfect farm will soon present itself.

Urban and suburban farming

You don't have to live in a rural area to be a market farmer. Many small-scale growers farm in the city or suburbs. Some piece together farms from multiple backyards or urban lots. Many urban farmers don't live on the land they farm, which has some obvious drawbacks and some benefits. On the one hand, you can't be as vigilant about your farm when you aren't there 24/7. You can't defend your crops and livestock from deer or dogs, monitor your greenhouse temperatures at night, make permanent capital improvements, or work every waking moment. No, wait . . . that may be one of the benefits of not living where you farm. I have known a few market gardeners who choose to live in town for all the benefits city living confers, including saving gas, less time taking kids to activities, more social interaction, more cultural activities, and so on. For them, farming is like a job that they come to every morning and leave every evening.

Urban farming comes with its own set of issues, including the potential for soil contamination, zoning and other regulatory problems, security and theft, and many others. Because there are so many considerations for an urban farmer, I suggest that you read one of the recommended books in the resource list before deciding whether to start farming in a city.

Incubator farms

In a few places around the United States, nonprofit organizations have created opportunities for beginning farmers to have free or low-cost access to land and equipment. Known as incubator farms,

Clearinghouse of Free Information

Programs to assist small farmers have proliferated in recent years, to the extent that it's impossible to identify them all in this book. Fortunately, there is a single website that is collecting information about all the free resources out there: Start2Farm.gov from the USDA's National Agricultural Library. It has sections for aspiring farmers and for those who are already farming. Organizations are invited to list their programs on the site, and it's free for all.

they are great for growers who have some experience and are ready to strike out on their own, but perhaps don't have the money to buy land and equipment. In general, an incubator farm hosts multiple growers who each run their own business but share the farm's facilities and tools. The original incubator, and the one most other programs look to for guidance, is the Intervale Center Farm Incubator in Burlington, Vermont. Many others are being started thanks to funding from the US Department of Agriculture's (USDA) Beginner Farmer/Rancher Development Program. In some cases, private landowners have created incubator farms. There is a National Incubator Farm Training Initiative that is compiling information about all the incubator farms in the United States. It's housed at Tufts University.

Renting land

In many places, land is too expensive to be financially viable for market farming. You simply cannot make enough money selling vegetables to justify the cost of a land purchase. Renting land may be a better option, at least in the beginning. There is plenty of land available for market farming; many landowners hold onto land as an investment, with the idea that they will sell it for retirement income or pass it down to their heirs. In the meantime, they may want to have the land farmed for many reasons—to keep their agricultural property tax exemption (see chapter 8) or to keep it from becoming infested by weeds and trees, for example. Your challenge is to find a desirable piece of land, then approach the landowner about renting it.

Before you do, you need to know the range of possibilities and costs for renting land. Tenancy arrangements vary widely. I know beginning farmers who "rent" land basically for free or in exchange for minimal work, such as mowing. Some exchange vegetables for use of land. At the other end of the spectrum, some pay thousands of dollars a year per acre plus a percentage of sales. It's really hard to predict what terms you might be able to negotiate, so let's take a look at traditional land lease arrangements as the basic way of looking at renting.

In general, there are three typical approaches: **cash rent**, which is a specific dollar amount per acre; **crop share**, in which the landowner and tenant share revenue; and **flexible cash rent**, in which the tenant pays a specific dollar amount per acre plus a percentage of revenue. (See the profile of Chip and Susan Planck earlier in this chapter.)

Cash rent is most advantageous to the landowner, who will receive payment regardless of the success of the farmer. But it can also help the farmer by making rent a fixed, predictable cost every year. With a crop share arrangement, the landowner shares the risk with the farmer. Flexible cash rent is a combination of the two.

Calculating a fair rent price is no simple matter; it's the subject of endless discussions and publications nationwide every year, based on farmland values and crop prices. In fact, the USDA's National Agricultural Statistics Service conducts the Cash Rents Survey annually in every state except Alaska to keep tabs on farmland values and rental rates. The survey applies mostly to large, conventional crop and livestock farms, but it's nevertheless instructive to know what a landowner might earn in cash rent from that kind of use of his or her land. In recent years, cash rents around the country have varied from less than $100 per acre in upstate New York to more than $1,000 an acre in California. You can find out the range in your state by visiting www.nass.usda.gov and searching for "Cash Rents Survey."

Once you know that figure, think of it only as the broadest of guidelines. The best arrangement depends on the grower's skill, the landowner's financial needs, and many other factors. The goal in any agreement is to make it a win-win for both parties.

Price is only the first and most obvious consideration in renting land. You should also think hard about the length of the rental term, renewal provisions, who will pay for improvements, liability insurance, and other issues. And although many people have oral agreements, you are strongly encouraged to have a written contract when renting farmland. Please read *The Landowners' Guide to Sustainable Farm Leasing* from Drake University's Agricultural Law Center; details are in the resource list at the end of the book.

➧ TURNING YOUR HOME ◄ INTO A BUSINESS

Perhaps you already live on the land where you want to start your farm. Realize that a farm is a business, and you should carefully think through the implications of turning your home into a business. What do you visualize for your property? Do you want to keep it private and take your crops to town? Do you want to start a pick-your-own or retail farm store and invite the public? Or maybe you can see taking the middle course, of having a CSA with pickup by members at your farm. The next chapter, on the markets you might serve, will help you clarify your goals.

Before you plow your fields and put up a greenhouse, be sure your property is zoned to accommodate your plans. Many rural counties are flexible about land use, but those that are near cities often have more stringent zoning laws. The county planning department should have land use maps that show the zoning category for your land. Once you know how it's zoned, you can ask to see the restrictions for that type of zoning. You may find that you can do whatever you want. Or you may find that you can grow crops but not sell them from the farm, or that you can sell your own crops at a farm stand but not such purchased products as potting mix. Zoning regulations may allow you to build a hoophouse in the field but not a permanent greenhouse.

If the current regulations don't permit the uses you want to incorporate into your farm, you can request a variance. You would be wise to talk to a lot of people about your plans and gauge their reactions. You want to know which way the wind is blowing before you make a formal request for a variance. You also want to know the history of requests such as yours, so that you can anticipate problems and try to either address or sidestep them.

Going Organic

Another decision you should make when you're starting out is whether you want to get certified organic. The reason it's important to think about this early is because it may take up to three years for you to get certified, and you don't want to base your business plan on a set of assumptions that won't be true for three years.

"Organic" is a label regulated by the US Department of Agriculture. Farmers who want to call their products "organic" have to get certified by a USDA-accredited certification agency. The only exception is the very small grower, with less than $5,000 in farm sales per year. And even those small organic growers who are exempt from certification must follow organic rules established by the National Organic Program. Land cannot be certified until it has been free of prohibited substances for three years.

The rules for being organic are extremely comprehensive and cover everything from the seed you buy to the boxes you use to take your crops to market. The organic standards require you to use certified organic seeds and plants, unless you can prove they are not available for the varieties you grow. They control all production inputs, including potting mix in the greenhouse, fertilizers, pest control products, and postharvest products. They require you to keep extensive records of everything you do to your crops and of every product you purchase.

Some growers chafe under this regulation, but most find little to object to. For the most part, organic standards are consistent with best management practices recommended for small-scale producers who are concerned about the environment and human health. A few standards—the organic seed and plant requirement being one—are a little difficult to deal with. And the record-keeping can be time-consuming.

The cost of certification is another obstacle. Being certified organic can cost anywhere from a few hundred dollars to several thousand dollars per year. There are dozens of accredited certification agencies, each with its own fee structure. The biggest cost is for the inspection, which is required annually. If there are other organic farmers in your area, it makes sense to use the same certification agency so you can schedule inspection visits together to save on travel costs for the inspector.

There have also been some federal funds made available in recent years to help growers pay for certification. The program will pay up to 70 percent of the cost of certification, to a maximum of $500 per year. At this writing, only a few states have remaining funds from the program, but there is always the possibility of renewed funding in the future. Certification agencies will know, or you can do an Internet search for "National Organic Cost-Share Program" and the name of your state. The best place to familiarize yourself with the process of organic certification is on the USDA's National Organic Program website: www.ams.usda.gov/nop. There you will find the standards themselves—all 554 pages of them. You will also find what's called "the National List" of products you can use in organic production. And you'll find state contacts for the cost-share information in the states where it is available.

Farm Tours

The single best thing you can do when starting out in farming is to visit other farms. You can learn a great deal just by being on other farms and seeing how things are done. I recommend that you visit every farm you can in your area, and that you consider a few working vacations to visit farms in other states. The combination of learning what works locally and seeing how farms operate elsewhere will give you a valuable perspective on how to mold your own business.

Visiting a farm isn't as simple as dropping in on another farmer, though. Most farmers are much too busy to be able to entertain visitors during the growing season, which is when you want to go. Some worry about sharing their trade secrets with potential competitors. So don't start calling farmers to request a tour, unless you already have forged a good relationship with them and know that they want to help you get started.

A better approach is to get connected with the sustainable agriculture and horticulture groups that hold field days on successful farms. Locally, contact your Extension office and other organizations to find out if any field days are scheduled. Usually these are "twilight tours" held in the evening during the summer. Often a tour will visit several farms in one evening. Check the websites of the organizations listed in the resources to see about visiting farms in other places, too. Many of these groups have educational programs that include farm tours.

Make every effort to get out to these events because not only will you see the farm and hear from the farmer, but you also will benefit from the experience of the other people in attendance by paying attention to the kinds of questions they ask.

You may find that you are the only new farmer, and feel embarrassed to ask questions. If that's the case, ask the farmer for his or her email address and permission to send follow-up questions.

Take a camera and take lots of photos. Take wide shots and close-ups. It doesn't matter whether the photo seems worthy at the moment. As soon as you get home from the tour, you'll probably think of a dozen questions you should have asked. If you have photos of the place, you may be able to find answers there.

The website will also provide you with a list of certifying agencies, which at this writing numbers 50 in the United States. The USDA lists them by the states where they are headquartered, not by the states where they conduct certifications. If you don't see your state on the list, you may have to check around locally to find out what agencies are serving your state. Some certifying agents say they are national, but be sure to comparison-shop on fees. One of the effects of the federalization of organics is that regional or state-based groups that used to certify farms and provide educational programs are no longer allowed to do both. Some have divided into two separate legal entities; others have chosen to be one or the other to their members. In any case, the best source of information on being certified will be these smaller membership groups, most of which hold conferences and farm tours and publish newsletters. If you haven't already done so, get connected with one of these organic groups.

➣ OTHER LABELING PROGRAMS ◂

Many small-scale growers don't see the need for organic certification, even if they use organic methods of production. In some cases, the expense of certification is a deterrent. Others don't want to bother with the paperwork, and some farmers just have philosophical objections to participating in a federal program.

Whatever their reasons, noncertified farmers have come up with alternative ways to describe and label their products, such as "Beyond Organic" or "Ecologically Grown." These labels may require specific land or infrastructure modifications, so it's helpful to know about them when first setting up your farm.

One such program is Certified Naturally Grown (CNG), which is a web-based, low-cost certification system in which the grower agrees to follow the National Organic Program rules. Inspections are done locally, by other farmers, Extension agents, even customers. Although it is basically an honor system, each year a certain percentage of CNG farms are randomly selected for pesticide residue testing of produce, at no cost to the farmer. At this writing, about 700 farms and apiaries are certified under the program. An annual financial contribution of at least $110 is required for certification. For more information, visit www.naturallygrown.org.

Another sustainability certification program that does not require compliance with national organic standards is the Food Alliance. Standards for crop production allow limited pesticide use in the context of an integrated pest management program. See foodalliance.org.

In addition, many local groups have their own labeling programs, such as the "Buy Fresh, Buy Local" campaign, and some environmental organizations will certify farms that comply with certain standards (for example, "Salmon Safe"). Such "eco-label" programs are proliferating despite concerns that they are confusing to consumers. Consider whether a label will be helpful in explaining your practices, then choose the program that fits your needs.

Picking a Good Name
for Your Farm

Thomas Jefferson had Monticello. David Wallace has Mr. Green-bean's Market Garden. Louise Rickard has Oh, Sweetpea! Thomas Fister has Double Stink Hog Farm. My farm is Wild Onion Farm.

Every farm needs a name. What's in a name? Quite a bit. A farm name speaks volumes about its owners—about their products, their dreams, their personalities, their sense of humor. The tradition of naming farms, which dates to an age when there were no addresses, continues today for reasons just as important.

A farm name gives the farm's products an identity. It gives customers an easy tag to remember the farmer by. And it should convey a positive image that makes the customer feel good about spending money with that farmer.

By far the most common approach to naming a farm is to use the owner's name: Peterson's Orchard, Fields' Organic Farm, Jackson Family Farms, etc. There's a lot to be said for using a family name—it ties the owner, and the owner's reputation, to the farm products. That can be helpful for a family well known in the area where they sell. It seems to be limited, however, to people with names that are easy to spell and pronounce.

Another common tack is to name the farm for a geographical landmark, such as the road it's on, the creek that runs beside it, or the town nearest to it. That's a good strategy, too, because it helps customers remember location as well as name.

Farmers often use some kind of natural element from the farm, such as the flora, the fauna, and the topography. The distribution of the cedar tree is obvious from a survey of small-farm names: Cedar Brook, Cedar Creek, Cedar Crest, and Cedarville, from Massachusetts to Washington. Meadowlarks and blue herons are equally widely distributed, with farms of those names in Idaho, Michigan, New York, Oregon, Washington, and Wisconsin. Hillside, Creekside, River Bend, and Long View all speak of the farm's layout.

Many farm names are designed to evoke a pastoral image. The words "fair" and "green" help here. Fairmeadow Farm, Fairplain Farm, and Fairview Gardens were all employed by

Branding Your Farm

Arnosky Family Farms is a cut flower and vegetable farm in Blanco, Texas. The owners, Frank and Pamela Arnosky, chose "local color" and a shade of blue to brand their farm. They painted their barn the bright color and got a professionally designed logo that they use on marketing materials, including the label on all their flower bunches.

As important as your farm name is, it's only part of the identity you will begin to create for your business. You should also start thinking about a logo, a website, signs, and other printed materials. All of these materials should work together to build a brand for your farm. Branding is not just for big companies; it's for any business needing a coherent way of presenting itself to the world.

Think carefully about the image you want to project. Do you want to be seen as a young, hip farmer—or maybe as a deeply rooted member of your community? Use your name, logo, colors, and motto consistently in all marketing materials to create a business image that truly represents your farm.

Logos run the gamut from hand-drawn to professionally produced. A logo can cost from nothing, if you do it yourself, to several thousand dollars, if you hire a pro. A middle ground for many growers is to hire a student with artistic ability and graphics training and pay him or her a couple of hundred dollars. Another is to use a crowdsourcing website, such as crowdspring.com or 99designs.com. You describe the product you want, such as a logo or brochure, and state how much you're willing to pay. Designers pitch ideas, and you select the one you like best.

farmers who have written books about their farms. In the "green" category: Acres, Earth, Garden, Meadows, Mountain, Pastures, Ridge, and Thumb. "Sun," "moon," and "wind" all help convey natural images, too.

Some farm names have a rich story behind them. Janette Ryan-Busch in Iowa City, Iowa, was leaning toward something mystical because her children kept telling her there were fairies and gnomes in the woods.

"I didn't want it to be really blatant," she says. "You don't want to say "trolls" or "gnomes" or "fairies" because then people would think you're weird."

However you get your logo, be sure you have it in a format that is accessible to you on your computer so that you can use it whenever you need it. Ask the designer to produce several versions with different resolutions: 72 dots per inch (dpi) for a website and 300 dpi for printed materials. If the logo is in color, get a black-and-white version, as well, for letterhead or newsletters. Get a version with a transparent background, too, for screenprinting on products.

You can get your farm name and logo enlarged onto a banner at most copy shops. Banners and other signs are important to cultivating repeat business. Many customers won't remember what *you* look like, but they will remember your logo or farm name, so give them a way to find you in the future. You can also have your logo embroidered or imprinted on hats, t-shirts, and aprons to wear at market. Some growers buy inexpensive pens printed with their farm name and hand them around liberally. There are hundreds of companies that provide custom-printed pens and other products. Check with your local office supply store, or do a web search for "promotional products."

Websites have become so routine in the business world that you run the risk of being considered a dinosaur if your customers can't find you on the web. The best websites are those with multiple pages and many beautiful photos. But even a single homepage that tells customers where to buy your products is helpful. If you don't know anything about web design, you can hire help—again, ranging from a few hundred dollars for a high school kid to thousands for a web design firm. Or you can buy a book about creating a website and do it yourself. It is not, as they say, rocket science, but it does take time—the one thing that is always in short supply for market farmers.

You can get ideas from other farms' websites. Start with a directory such as www.localharvest.org, which gives farms a free listing; if the farm has a website, you'll find a link to it. You'll also want to get your farm listed on Local Harvest as soon as you are in business.

She hit upon Fae Ridge, "fae" being an archaic English word for wood sprites. Most people never even ask her the meaning of Fae Ridge Farm.

"The people who know what that spelling means are already okay; they don't think you're weird for believing in wood sprites," she explains.

Some farm names reveal the aspirations of their owners: Someday Farm, High Hopes Homestead, Grateful Farm. Others focus on the hardships of farming: Agony Acres and Achin Back Farm. Others have a spiritual tone, such as Guided Path Farm, Infinite Garden Farm, New Harvest Farm, and New Morning Farm.

Then there's the pun put to good use for the sake of a farm product. Let Us Alone is a specialty lettuce farm in Florida; ditto for Let Us Farm in California. Herb farms do particularly well with this motif, thanks to the double entendre of "thyme." How about a wild time on Wild Thyme Farm?

David Wallace has a knack for wordplay. Mr. Greenbean's Market Garden near Kalispell, Montana, came to him because of his fondness for green beans. He has another business in winter removing nuisance animals like skunks. He calls it Critter Ridder.

Louise Rickard of Bristol, Vermont, started her flower business as a way to stay home with her baby, whom she often called "Sweet-pea." So she named the business after him—Oh, Sweetpea! People frequently comment that they love her name, she says. And it lends itself to beautiful artwork.

"I really don't sell a lot of sweet peas, but I can't quit growing them because they're my namesake," she says.

Descriptive names are good for business, too. You know what you can buy at Everyday Bouquets, Herb's Herbs, and Specialty Produce.

Some names are just nice; they roll off the tongue and are pleasant to the ear. Thousand Flower Farm, Lazy Lightning Farm, Earthly Delights, and Bulrush Farms are good examples.

The most unusual farm name I've encountered was Double Stink Hog Farm in Georgetown, Kentucky. Thomas Fister, the owner, did indeed raise hogs. He also sold bedding plants, nursery stock, sweet corn, summer vegetables, and pumpkins from an on-farm store. Oh, and hats. Every year, he sold a couple hundred hats with his farm's name on them.

"I was told by some that our business name should be Tom Fis-ter Farm," the owner said. "They told us that Double Stink wouldn't work. But it was so catchy, everyone in town knows where Double Stink Hog Farm is. The TV stations have really played it up, especially in the fall of the year, when we have our pick-your-own pumpkins. They fight over which one will get here first to do a story about us."

Double Stink Hog Farm is now defunct, but its name will be remembered.

CHAPTER 2

The Markets

What will you grow, and where will you sell it? You can approach this question from two different angles: Grow what you enjoy and then look for a market. Or identify the best market opportunities in your area, and then choose crops that are in demand. Either strategy will work, and most market farmers employ a combination of the two. Start with what you can grow really well so that you have a high-quality product to take to market, then look for other niches in the markets where you succeed.

Let's begin by exploring the opportunities. Growers of high-quality produce and flowers will find many potential markets for their products. In many places, growers find that no matter how much they grow, demand exceeds supply. They are the lucky ones who can pick and choose their buyers. Most growers have to sell in multiple channels to be able to move all their products. In any case, multiple channels exist; in this chapter, you'll learn more details about each of them so you can decide where to focus your efforts when getting started.

Farmers Markets

The most important market opportunity for small-scale and beginning growers is the farmers market. For just a few hundred dollars a year, you can essentially open up shop in a prime location with literally thousands of customers coming to your door to buy your products. There is nothing like it in any other kind of retail business. And though the existence of a farmers market in your local area won't guarantee success, it certainly is a great advantage for newcomers to direct marketing.

National Count of Farmers Market Directory Listings

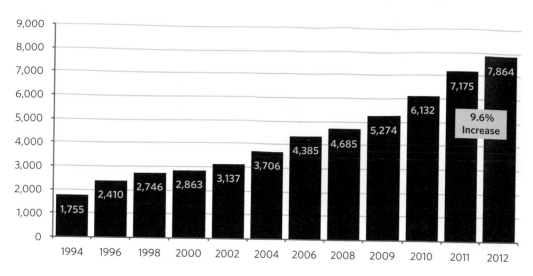

SOURCE: USDA-AMS-Marketing Services Division

When the USDA started collecting information about farmers markets in 1994, there were 1,755 markets in the United States. By 2012, the number had jumped to more than 7,800. Those figures are subject to some debate because collecting the data is difficult given the independent nature and widespread distribution of markets throughout the country. Nevertheless, there is no question that the United States has experienced an explosion of farmers markets in recent years.

The USDA has not tallied the number of farmers who sell at these markets, nor the amount of revenue generated, so it's difficult to assess the total economic value of farmers markets nationwide. However, a study of farmers markets in Iowa in 2004 provides some enlightening data:

❖ 1,600 vendors sold at 189 markets. With about 2.9 million residents, Iowa has the greatest number of farmers markets per capita in the nation.

❖ Total direct sales at farmers markets were estimated at $20 million, for an average of $12,500 per vendor. Nearly half of all sales were from fruits and vegetables, an additional 21 percent from baked goods.

❖ 72 percent of sales occurred at five markets in urban areas with more than 68,000 population.

The organization Local Food Shift asks shoppers at the Boulder (Colorado) Farmers Market to pledge to spend 10 percent of their food dollars on local food. The organization calculated that 25 percent "food localization" would give Colorado 31,022 new jobs and $1.3 billion per year in additional wages. Learn more at localfoodshift.org.

❖ 55,000 people—less than 2 percent of the state's population—attended market once a week.

The data for a small state like Iowa make it obvious that there is enormous potential for farmers markets in the future. The growing popularity of local food among consumers, coupled with the astronomic increase in the numbers of markets, means that this is a very good time to be getting into farmers market sales.

❧ WHO WILL SUCCEED ☙

An accomplished grower selling at a successful farmers market can make several thousand dollars per market. The most successful markets are in urban areas and have dozens of vendors and thousands of customers. Many growers at those markets can virtually sell out every week, so sales depend on their production capacity. At smaller markets, the usual sales figures are less than $1,000 per market, sometimes only a few hundred dollars. Beginning growers may sell less than $100 worth of items, unfortunately, largely because they haven't built up a customer base yet.

Selling at farmers markets takes skill and perseverance. First of all, you must grow significant amounts of high-quality produce, with a lot of choice for your customers. You must handle and display

A vendor at the Fulton Street Farmers Market in Grand Rapids, Michigan, greets customers with a friendly smile and an abundant display.

your produce as well as a supermarket does, because that's what your customers expect. You must price it appropriately so that you make a profit; don't ever make the mistake of thinking that you can get ahead by selling at lower prices than everyone else. You must have a pleasant personality that attracts customers rather than frightening them away. You must be willing to sell yourself and your products by being enthusiastic and knowledgeable.

Here's a question I like to ask vendors when I'm a shopper at a market: "I'd like to buy tomatoes for sandwiches (or eggplant for grilling, or whatever). Can you advise?" That's an open-ended question, and it sometimes takes people by surprise. But it's a legitimate question, especially if the vendor has several kinds, sizes, and prices of the item I'm interested in with no signage explaining the differences (more later on the importance of signs). The good marketer will show enthusiasm for the question and will name variety names and tell me about flavor. The bad marketer won't know. Guess which vendor gets my money?

There is definitely a personality type that is well suited to selling at farmers markets: gregarious, cheerful, accepting of all types of people, and proud of his or her work and products. If you don't fit that

description, you may be deeply unhappy going to farmers markets—
and that is going to decrease your chances of success. If you would
rather be home farming than at the farmers market selling, you might
want to think about a different venue for selling your products.

❧ GETTING STARTED AT MARKET ❧

If you decide you want to sell at a farmers market, your first job
is to survey all the markets within a reasonable driving distance.
Although it may be easiest to sell at a small market in your own
town, you might be much better off driving an hour or more to a
bigger city. So find out where and when all the markets are held,
and start visiting them. Go see what is being sold, what prices are
like, and whether there are any obvious gaps you could fill. Find
the market manager, introduce yourself as a potential vendor, and
ask where to read the market rules. Find out if there's a waiting list
and how long it is, or whether vendors are selected based on what
they grow. Find out if it's a producer-only market (which is generally
much better for growers) or if vendors are allowed to resell produce.
Find out what it costs to sell there.

Every market has its own rules and its own personality. Do your
research early and you are more likely to find a market where your
products will be popular and get a good price.

Well-established markets may be hard to get into as a new vendor. Popular markets tend to have waiting lists because great sales
income keeps vendors coming back year after year. If you live near a
thriving market, apply right away, even if there is no room currently
available. You can get your name on the waiting list in any case. At
some markets, vendors are selected based on their product offerings, and you never know when a market will want your products to
increase the market's overall diversity.

If you don't get accepted right away, there are many new, smaller
markets that are begging for vendors. One of the problems veteran
growers have identified in recent years is the proliferation of farmers
markets within a given geographic area. For example, Kansas City
had 15 markets a week in 2005; seven years later, there were 48.
Most other metro areas have experienced similar growth. Having
multiple markets in the same city or town can dilute attendance at
established markets, and it can force some growers to attend more
markets just to stay even with previous sales. After you've sold at a
few markets and realize just how much work is involved, you will

understand why you need either to limit the number of markets you attend or hire staff to attend for you.

Don't assume that you can't succeed at a particular market just because there will be a lot of competition. I can tell you from experience that a flower grower will do better at a market where there are already a couple of flower growers than at one where no one has ever sold flowers before. The same is true with organic produce, or watermelons, or virtually any crop. If you are the first ever to sell a particular item, you may spend most of your first season just letting people know it's available. However, you should always be attuned to opportunities to sell products not currently offered at your markets.

Chapter 7 will provide more advice about the art of selling at farmers markets.

Farm Markets

For some farmers, the best place to market is right on the farm. A retail market can be a convenient and fun way to sell your products, for the right kind of person in the right location. The best on-farm marketers don't worry too much about their family's privacy, and they don't mind having strangers watching them at work. They are tolerant of obnoxious customers, such as the ones who come up your driveway even when the sign says you're closed or who lay on the horn when your gate is locked. The best markets also are in a location that is convenient for the public, either on a paved road or not far off one, close enough to a population center or tourist traffic to draw sufficient customers.

An on-farm market can run the gamut from a seasonal affair, such as a pumpkin patch in fall, to a year-round market with a huge array of products for sale.

Whatever the scale, on-farm markets create an entirely new layer of work for the farmer. When you have a market, you really are a storekeeper, with all the additional responsibilities of buying products, keeping inventory, ensuring freshness, complying with labor and tax laws, making your building handicapped-accessible, maintaining displays, creating signs, and so on. Growing the product becomes secondary to marketing it, and many farm marketers eventually decide it's easier to buy produce from other farms than to grow it themselves.

The Chimacum Corner Farmstand on the Olympic Peninsula of Washington is a rural grocery store that sells food from many small local producers.

Although you may think the most important thing about your market will be the inviting atmosphere and creative displays of excellent produce, experienced marketers will tell you that equally high on the list of attractions are parking and restrooms. If you don't provide good parking and clean bathrooms, all your other attributes may go unnoticed.

A farm market can be a good job for a farm family member who is more interested in marketing than growing. Or, on a small scale, you might consider an honor-pay stand. Many growers who have sold that way say they rarely have problems with people stealing from the cash box or taking food without paying. Put up clear signs with prices, bolt a payment box to the wall, and empty it every few hours. An honor-pay stand may not provide enough revenue to support you, but it can be a nice addition to your other sales, without a lot of extra work.

If you are considering creating a retail market, a great resource will be the North American Farmers Direct Marketing Association (NAFDMA). You can become a member and plan to attend the annual conference, which usually begins or ends with a tour of farm markets. Visiting farm markets with a group of experienced farm marketers is the best possible education. Not only will you see the host markets, you'll also hear the conversations of people who are doing similar work all over the country. Many NAFDMA members are involved in entertainment farming ventures, so you'll find plenty of ideas about this type of business. Contact information is www.nafdma.com or 413-529-0386. Two business matters you should tend to before you do anything pertaining to an on-farm market: First, talk to your insurance agent about your plans to be sure you have adequate liability coverage. And second, talk to the zoning department in your local government to be sure you will comply with all the regulations. You'll read more on these matters in chapter 8.

Agritourism

Agritourism and agritainment are two terms that refer to farm-related activities that bring visitors to the farm. Agritourism can refer to something as simple as a U-pick farm or as complicated as a farm bed-and-breakfast. They can be designed to appeal to local folks or to tourists from afar. There is growing interest in visiting farms; people want to get closer to the source of their food, to learn how it's grown and to share that with their children. Also, theme vacations are increasingly popular—hence the growth of ecotourism, culinary tourism, and eno- (wine) tourism.

In all cases, agritourism is not cheap. You need to spend money on staff to supervise visitors, especially children. You have to spend on maintenance to make the farm attractive and safe for visitors. You need more expensive liability insurance when you're having customers on the farm, and you may need to invest in infrastructure such as restrooms, buildings, signs, and parking lots. That's why most destination farms offer lots of activities; once customers come out to the farm to shop, you want to have a wide range of products to sell to them.

The options for agritourism are diverse. Here are some of the possibilities:

U-PICK

U-pick farms are a natural extension of the on-farm market—you can have products ready to purchase in the shop, and you can offer your customers the option of going outside to pick it themselves for a lower price.

After a period of decline when people seemed too busy to pick fruits or vegetables themselves, U-pick farms are experiencing a comeback. Today's consumers may be looking for a bargain because of tough financial conditions. Or they may see U-pick farms as a form of inexpensive entertainment or education for their kids. In any case, U-picks used to be the domain of berry growers, but recently vegetable and cut flower farmers have adopted the model with success.

In interviews with growers all over the United States and Canada, no easy guidelines emerged to help you assess the potential for pick-your-own (PYO) success in your area. However, here are some trends that might work well for your farm.

Cut-your-own flowers

U-pick flowers can be a great addition to a farm stand or they can be a stand-alone attraction. The idea is still novel enough that it gets free press attention and can be sold as an entertainment opportunity for families, bridal parties, and other social groups. Most U-pick farms find it's easiest to grow a wide variety of flowers, then set price points based on the size of the bouquet (number of stems or filling a certain container size). Another approach, somewhat more complicated but possibly more profitable, is to grow flowers in groupings based on their prices. For example, you can plant all the annuals in one block and post a sign indicating they are 50 cents per stem. Another block might have the valuable perennials, such as peonies and delphiniums, priced at $2 per stem. Clear signs are essential in this kind of setup, because you don't want customers to balk at the cash register, saying they didn't understand the pricing. Weatherproof signs should be erected in each price block, identifying the flowers by name and price.

Grass paths between the flower beds make for a more pleasurable experience for customers and reduce the risk of people walking on the beds. Some flowers, though, should only be cut by a professional (you) and can be sold at the cash register as add-ons. Examples of these would be lisianthus grown on support nets, lilies that you want to perennialize and therefore don't want the stems cut too long, and tulips that you want pulled out rather than cut.

Fruits and vegetables

Some of the most popular crops for U-pick are asparagus, strawberries, raspberries, blackberries, blueberries, apples, grapes, tomatoes, and pumpkins. Some farms offer a wide array of vegetables, including peppers, squash, cucumbers, green beans, okra, and more. Many farms also offer ready-picked produce at their farm stands, and customers seem willing to pay a significantly higher price when they consider the labor they would have to expend to get the lower U-pick price. Some farms charge double for ready-picked produce. When the weather is pleasant, more people are willing to pick their own versus when it's cold or hot outside.

A labor-saving strategy used by some farms is to invite customers to pick for free; the catch is that they have to leave half (or two-thirds) of what they pick with the farm. In that way, the customer gets free food and the farmer gets free labor to pick produce that he or she can sell in the store at a higher price. This is an effective strategy when you have a glut of a particular crop that is time-consuming to pick. Blueberries and green beans are good examples.

Problems with U-pick

Whatever the crop, U-pick operators need to be mindful of liability issues and be sure they have adequate insurance coverage. You have to lay out your fields to be as safe as possible for customers and walk them often to remove any hazards that may have been brought in or blown in. Firm paths, clear signs, appropriate harvest containers, and supervision will minimize the possibility that someone will get hurt.

U-pick farms also must take measures to prevent customers from ruining the crop. You must direct people to the field where you want them to pick, provide clear instructions on how to pick, provide suitable containers, and make your prices clear. One of the biggest problems many U-pick operators face is customers who eat as they work, often consuming twice what they eventually pay for. Some sampling is inevitable, but a sign at the entrance to the U-pick asking people to weigh and pay before they eat will make most people aware that you expect and deserve to be paid for everything they pick.

❧ ENTERTAINMENT FARMING ❧

Some farmers love to have people out on their farms, and they create many opportunities for fun and interesting outings. Unless your farm is a nonprofit with an educational mission, the goal of any activity should be increased revenue. So think about the ways an activity can lead to sales, whether it's by a fee or simply by making your products visible and easy to purchase. The smart marketer offers multiple activities and ways to spend money at the farm.

A sign at Pond Hill Farm in Michigan lists the many ways visitors can enjoy a visit to the farm.

Farm activities are limited only by your creativity; here are some of the more common ones:

Pumpkin patches are a long-standing American tradition popular with families and elementary school classes. Kids love to run through a field of pumpkins to pick their own. Although many pumpkin farmers grow their own, it's not uncommon for them to also buy pumpkins and spread them out in the field. Big pumpkins, ornamental corn, gourds, and other products should be offered at the farm store to increase the average sale.

Corn mazes can be a profitable business for farmers who have a good climate and soils for growing corn plus a big population base nearby. Growing a corn maze is much different from growing a field of corn for harvest. Variety selection is important to ensure the corn will be tall enough (7 to 9 feet, usually) and stand through the fall, when most mazes operate. The biggest attention-getters in the maze world are those cut into a design visible from the air. Several companies offer maze-making services, including designs and training in how to cut the maze, and a few even will do the work on-site. Rutgers University and Purdue University have publications about growing corn mazes.

Pumpkin cannons or other devices for chucking vegetables at a target are offered on many farms. Be sure to talk to your insurance agent about this idea before you make one—because of the higher liability risk, the insurance company may not want to cover it.

❧ FOOD SERVICE AND LODGING ❧

Although most farmers can't imagine adding another venture as exhausting as food service to their already-full plates, hospitality businesses can be operated on-farm by another family member or partner.

Farm-to-fork dinners are a growing trend. The general format is for a local chef to prepare a gourmet meal from farm ingredients and serve it on a long table in the field, orchard, barn, or greenhouse. In some cases, farm dinners are a onetime event for a farm, organized by an outside company. For example, Outstanding In The Field is a California-based business that has traveled around the country since 1999, hosting dinners at dozens of farms each year. The company's mission is "to re-connect diners to the land and the origins of their food, and to honor the local farmers and food artisans who cultivate it." A typical event costs $180 per person, and includes a gourmet meal, wine, a farm tour, and talks by the host farmers. A long table for as many as 200 people is set up outside.

Farm dinners also can be a great way to raise funds for nonprofit organizations, such as a farmers market or sustainable agriculture group. And, increasingly, they are run entirely by the farm owners and held regularly throughout the warm season.

Farm restaurants are unusual, but they do occur on working farms. Fresh ingredients, seasonal menus, and a clean, inviting dining space are important elements of a successful on-farm cafe. Farmers are advised to check with local health departments for any regulations or required permits before investing time and money into a food service business.

Huckleberry House is a farm stay adjacent to Finnriver Farm and Cidery in Chimacum, Washington. Guests can walk a self-guided farm trail to learn about organic vegetable and fruit production or visit the cidery's tasting room. Farm activities are completely optional, though, and some guests come just for the peace and quiet.

Farm stays, long established in Europe, have been popping up in the United States in recent years. Typically, the farm constructs or converts a cottage into guest quarters outfitted with high-end bathroom and kitchen fixtures, comfortable furniture, and luxury linens. Visitors, who pay more than they would at a local hotel, are free to participate in the farm activities to the extent they wish. Some people want to be fully involved and work side by side with the farmers; others just want a quiet retreat in the country where the air is clean and the only sounds are the chickens. You can find examples of all kinds of farms that put up overnight guests at www.farmstayus.com.

Wedding venues can be a summer sideline on some scenic farms, especially those that grow cut flowers. Some of the facilities that may be required include a dressing room, bathrooms, tents in case of rain, refrigeration for reception food, tables, and chairs.

Community-Supported Agriculture (CSA)

Community-supported agriculture is a system in which consumers connect directly with a farm and agree to purchase a certain amount of produce for a growing season. The exact structure of the

Sunset Dinner on a Vermont Farm

Farm dinners at the Schlosser's Vermont farm PHOTOGRAPH COURTESY OF SARA SCHLOSSER

After 25 years of selling at farmers markets, through a CSA, and to restaurants, Sara and Bob Schlosser of Sandiwood Farm in Wolcott, Vermont, made the leap into agritourism. They are hosting farm-to-fork sunset dinners on their farm and developing related opportunities for people to visit.

The new endeavor proved to be a lot more work than they expected, but it has reignited their excitement about farming.

"It is very rewarding having customers who have bought produce from us at the farmers market for 20 years come to our farm," Sara says. "I love connecting people, food, and farm."

Sara and Bob have had customers on their farm before, to purchase plants and maple syrup in spring, or for occasional events. They had planned to start offering buffet-style casual dinners using compostable plates and forks,

financial relationship is variable. Some CSA farms get paid for an entire season's worth of produce in advance or in two payments. Some CSAs get paid after delivery. However, the financial arrangement is only a small part of the overall relationship known as CSA.

Robyn Van En, the late cofounder of the first CSA project in the United States, used to tell this story when she encountered growers who were skeptical about it:

followed by an agriculture-themed movie. But their daughter, Sandi, a chef and graduate of the New England Culinary Institute, had a different vision. She wanted to cook gourmet meals and serve them on fine china and table linens. Their son, Kyle, a student at the University of Vermont, wanted to be involved in the events, too. So the family held its first dinner in 2012—just 20 guests at first, setting up borrowed tables in the center of a high tunnel between beds of greens and tomatoes. They charged $50 per person for the food and allowed guests to bring their own beer and wine.

Chef Sandi prepared a menu of local foods, most of them harvested right there on Sandiwood Farm: fried green tomatoes with goat cheese; cabbage roll with farm beef; pumpkin soup served in a mini baked pumpkin; farm-made spinach pasta; herb-roasted chicken and roasted root vegetables. The farm interns helped with all aspects, from growing and harvesting the produce that morning to setup, serving, and cleanup.

The events were so popular, with a waiting list each time, that the Schlossers soon doubled the guest list to 40 diners. Throughout the summer, they learned all the little details of putting on a dinner, like when to start the coffee and how to get the appetizer plates washed in time to be used for dessert. Encouraged, they decided to invest in the business, buying an event tent, china, and their own tables.

They also installed a certified kitchen, which Sandi plans to use to create a line of value-added products. Her family has long had a mail-order business for its maple syrup, so the new products will have a distribution channel immediately. She also plans to do catering and teach culinary classes. Several customers have already approached the family about hosting corporate events at the farm. Sara is a justice of the peace who has officiated at many weddings over the years, and she expects to develop the wedding end of the business on the farm as well.

Sara cautions that farm dinners and similar events require a huge amount of administrative work, emails back and forth with customers, receiving payment, hiring staff, and thousands of other details. She also says that hosting a farm tour and dinner requires them to keep the farm looking beautiful and the fields well weeded.

But she is gratified that her family's foray into agritourism is off to a great start, and especially that it has brought the next generation into the farming business.

"Opportunity abounds to expand," she says.

A New England market gardener wasn't interested when people in his community first came to him and asked him to start a CSA project. He was satisfied selling his produce to restaurants, whole-foods markets, and at his roadside stand. He couldn't really see the need for a CSA, where people put up money at the beginning of the season to get a share of whatever a farm produces.

The farmers at Oakhill Organics CSA in Oregon take a photo of every week's share. Here are shares from February, May, September, and November. PHOTOGRAPHS COURTESY OF KATIE KULLA

His customers were persistent, though, and eventually he agreed to offer shares to 12 families. It worked well that summer, so the next year he expanded to 24 shares. But he still wasn't sold on the idea of converting totally to a CSA.

That summer, the grower was run over by his tractor and was laid up for weeks with broken ribs. His restaurant, roadside stand, and grocery store customers vanished. But his CSA members came out to his farm and got the work done.

The moral of the story is that a CSA is more than just a new way to market produce. It's an entirely new economic system based on community, trust, and commitment.

"The CSA is what gets me out of bed in the morning," Van En once said. "It's the most positive, hopeful thing that is going on in this country today."

Robyn Van En died in 1997, but her work lives on in the more than 4,000 CSAs now operating in the United States. A national CSA center bearing her name has been established to help farmers start CSAs. The Robyn Van En Center at Wilson College in Pennsylvania can be accessed at www.csacenter.org.

CSA projects take many forms, but all have in common the mutual commitment of growers and consumers to the preservation of local agriculture. Consumers support a farm by sharing the costs of production and the risks of crop failures; growers share the harvest of fresh, pesticide-free food with the community. The price of a membership is usually referred to as the share price. Share prices around the country range from $300 to $600 for a six-month season.

Growers say the benefits of a CSA include getting a stable income, developing personal relationships with customers, contributing to the good of the community, and preserving farmland. Disadvantages for growers might include giving up control of some aspects of the job and sustaining the system over the long term.

"The whole idea of a CSA is transforming," says Dan Guenthner, the grower for Common Harvest Community Farm in Minneapolis. "I think growers need to be open to the possibility of really changing their lives."

The classic CSA, based on the European model, is one in which a group of consumers pays all costs of production, including the growers' salaries, benefits, insurance, and even land rent or mortgage. The farm budget is divided by the number of family-sized portions of food the growers expect to produce, to derive a cost per

CSA saved his farm. John Peterson was a failed farmer who lost his family's land in the farm crisis of the 1980s. He started Angelic Organics in 1990 and grew it into one of the largest CSAs in the country, with more than 1,200 share-holders in the greater Chicago area. His story is the subject of the film *The Real Dirt on Farmer John,* and his eloquent essays about farming are collected in the book *Glitter and Grease.* John created the Angelic Organics Learning Center as a nonprofit educational organization to help people build local food systems. His farm hosts thousands of visitors every year for educational and cultural programs. Photograph courtesy of Angelic Organics, www.angelicorganics.com

share. For example, if the production budget comes to $30,000, and the growers think they can produce enough to feed 100 families, each share will cost $300. The sharers pay in one lump sum or several payments at the start of the season then receive their produce every week. In this example, the sharers divide the total harvest, with none sold through other outlets.

Community-supported agriculture can take as many different forms as there are farms. Today, the farm that sells only to its CSA members is rare; most farms offer CSA as a component of their marketing, and also sell to restaurants, wholesalers, and at farmers markets. They may offer CSA shares based on quantities that are reasonable for two or four people, and price them somewhere close to retail prices.

Growers also can sell just a few shares until they decide whether a CSA is the system for them. Prices for shares can be based on retail prices for the expected production, set somewhere between retail prices for organic produce and retail prices for nonorganic.

Emily Oakley of Three Springs Farm sets up for market in Tulsa, Oklahoma. Emily and her partner, Mike Appel, offer a "farmers market CSA" in which customers pay up front, as in a traditional CSA, then debit their accounts when they make purchases at the farmers market. PHOTOGRAPH COURTESY OF MIKE APPEL

Another strategy that falls under the general umbrella of CSA is to sell credit that can be used at a farmers market. For example, a customer gives a farmer $200 before the season begins, and the farmer issues $200 or more in coupons that can be spent at the farmers market. The idea has merit for growers because they still get money up front to help with production costs. And it's a benefit to customers because they can choose the produce they want, use the coupons when it's convenient for them, and, in some cases, get more for their money. And of course they still develop a relationship with one farm and can feel good about supporting local agriculture.

❧ COOPERATIVE CSAS ☙

Some CSAs are run by groups of growers rather than individual farmers. I can attest to the benefits of working cooperatively, having been involved in one of the longest-running cooperative CSAs in the country. Early in our vegetable farming careers, my husband and I had a CSA with 80 families, which we enjoyed tremendously but also found to be highly stressful. We were constantly worried that we wouldn't have enough of a specific crop to fill all 80 shares, or that we wouldn't grow enough different kinds of crops to keep our shareholders satisfied. After a few years of that, we applied for and received a grant to establish a cooperative CSA with other organic growers in our area. With seven other farms, we created the Rolling Prairie Farmers Alliance, which has up to 300 shareholders each season. By working together, we were able to supply our members with a diverse selection of produce every week for a 26-week season. Some growers specialized in strawberries and raspberries, others in greens, others in tomatoes and melons. All members in the group had other markets for their produce such as a farmers market or restaurants, so they had other outlets if the CSA couldn't take all of their production in any given week. The logistics of the CSA were pretty complicated but not overly burdensome.

Relationships among the farmers are the most important factors to consider in setting up a cooperative CSA. Be sure you know one another, and especially that you know one another's level of quality and commitment. And write bylaws that will allow the group to solve problems as they arise: For example, have a rule about who decides whether produce is of sufficient quality for your CSA; have a rule establishing a trial period for new members; and have provisions for how the group can dismiss a grower who doesn't contribute enough.

➤ MORE THAN A BOTTOM LINE ◄

As important as money may be, most CSA advocates are more excited by the intangible rewards they have found. They take pride in their role of educating consumers about the food system and watching people develop ties to the land they may have never felt before.

In some CSAs, members who are willing to work on the farm or at distribution get a reduced price. Most CSAs also hold work days and festival days to get all the members together at one time, to get to know each other, as well as the farmers.

Guenthner says that when he delivers, a group of sharers will often be sitting on a front porch, talking about the weather and the progress of the crops. One year, seven families showed up at the farm, without being asked, to help cover the crops when an early frost threatened.

The sharers' deep interest in production means that many decisions once made only by the farmer have to be shared, which can be a problem for some growers who are used to their independence. "It's not for everybody," said Rod Shouldice of the Bio-Dynamic Association. "A lot of farmers would rather just be by themselves, thank you. There's a gateway into the program. A grower has to feel 'I'm tired of doing it the way I've been doing it. I want to feel a connection to the people I'm growing for.'"

Packing CSA Shares

When CSA first arrived in the United States, most farms offered members a box of food that they packed themselves, offering whatever was available that week. That's how Food Bank Farm in Hadley, Massachusetts, did it for the first two years—until farmer Michael Docter realized the farm was retaining only 55 percent of its membership from year to year.

People joined because they liked the concept, but some left after a season because they found that it just didn't meet the needs of their families. Greens were a particular source of discontent for members. Although some wanted to eat collards, kale, and other types of greens every week, most members felt inundated by greens, especially early and late in the season when other vegetables were more scarce.

During its third year, the Food Bank Farm CSA implemented a mix-and-match greens table. The farmers put all the collards, kale, lettuce, mizuna, arugula, radishes, etc. on one table and gave shareholders a plastic bag of a certain size and told them to fill the bag with whatever they wanted.

Members were thrilled with the new system because they could take what they wanted in the quantities they knew they would eat. The farmers were thrilled, too, because they saved thousands of dollars in labor to bunch and pack the greens. The savings there enabled them to spend money on growing larger crops of the most desirable vegetables, such as carrots.

Eventually, the farm went to all-bulk distribution. The crew puts all the produce out; when members arrive to pick up, they check a chalkboard that lists the maximum amount of each item they can take, if they want it. Congestion in the share room disappeared when the scales were removed, and customer satisfaction went way up.

The farm had to increase production of certain staple crops. And it often has excess of certain other crops—but those are donated to the food bank.

Although Food Bank Farm was one of the first to offer a free-choice CSA system, many growers have since stopped packing CSA shares and instead allow members to take what they want.

Many growers are delighted to be able to pass on much of the management work and concentrate on production. "The stressful thing for me was trying to market it all at the same time I was growing it," said one CSA grower in Colorado. "The CSA puts the fun and creativity back into growing."

❧ INTERNET ORDERING FOR CSA ❧

CSAs originally gave shareholders a portion of whatever crops they had, whether the members liked that food or not. That led to a level of discontent and turnover among members who felt they couldn't eat all the food presented to them. So farms have become more accommodating of differences in food preferences. Some farms offer an either/or system, as in "Take a bunch of carrots or a bunch of beets." Larger CSA farms have found that they can just put everything out with an upper limit on each item (e.g., "Take one bunch of kale, take three bunches of herbs"). Interestingly, when there is free choice, it all evens out most weeks and everyone goes home satisfied.

These types of choice systems work best when members come to the farm itself or to another indoor pickup place. Farmers can bring boxes of cleaned produce and let customers fill their own bags or baskets.

For those who are delivering onto a porch or other quick-stop location, letting people choose their own produce may not be a viable option. Those farmers must prepack the shares. Now there is an alternative that will allow those producers to give members more choice—an online ordering system. The way most ordering systems work is that customers receive an email telling them what's available each week. They can then log on to the farm's website and choose items and amounts within certain limits. Farmers who have gone to online ordering like it because it tells them exactly how much to pick and pack. Shareholders like it because they get only what they want. See the list of resources for software programs for CSA farmers.

A simple use of the Internet is to send out emails to regular customers offering new products or extras, such as canning tomatoes. Have customers reply to your email with orders that you can bring to the farmers market or add to their CSA boxes. Because it's not automatic, this kind of ordering system is best used only with small numbers of customers.

Restaurants

The most important thing you need to know about selling to restaurants is that good chefs want to do business with local growers. From the elite, expensive restaurants to the high-volume, family-style eateries, food establishments can benefit from a relationship with a local grower.

This is not to say that every restaurant in your community is going to want to buy from you. You may, in fact, have to do some educating before you win any sales. But you should approach each prospective customer with the confidence that you are part of a national trend and that you have something great to offer the restaurant. Locally grown food is probably the hottest trend today in the restaurant business, and you can help the restaurants in your community join the trend. Consider this:

- The Chefs' Collaborative is a national organization with more than 1,000 members dedicated to promoting sustainable cuisine with local, seasonal, and artisanal food.
- America's celebrity chefs—those who write books, have television shows, and own successful, creative restaurants—are supporters of local agriculture. Alice Waters, Deborah Madison, Charlie Trotter, Thomas Keller, Mark Miller—all of these chefs have helped make local produce popular.
- Even in smaller cities, local chefs conduct cooking demonstrations at farmers markets, again reinforcing the idea that the best ingredients are locally grown.

Not every restaurant is trying to be on the cutting edge of food preparation. Many just want to serve lots of customers good food at the lowest possible cost. You may find interest at either type of restaurant, depending on what you grow, the quantities you can guarantee, and how competitive you can make your prices. Let's consider each type of restaurant separately.

⇒ HIGH-PRICED RESTAURANTS ⇐

Trendy gourmet restaurants now exist almost everywhere, including small cities and tourist towns. Chefs at these places tend to regard themselves as artists, and are less price-conscious and more quality-conscious than cooks at other types of restaurants. Most small growers have had the best luck selling to these pricey, gourmet places because they can get the most money for the smallest amount of produce.

If you think that should be your strategy, you need to learn all you can about specialty produce. First, make sure you know the lingo. Study gourmet food magazines to see what kinds of vegetables and herbs are being used in the trendy places. You don't want to go in peddling plain old lettuce and tomatoes; you'll need, instead, to talk to the chef about mesclun and heirloom tomatoes. And you definitely don't want to mispronounce radicchio and frisee (rah-deek-ee-o and free-zay). Study the specialty seed catalogs to find out what varieties are hot. You may be amazed at how much sophisticated chefs already know about growing produce; don't be surprised if a chef asks for varieties by name. But don't be afraid to suggest alternatives, either. For example, a chef may ask you for Galia melons, because he or she has read about how wonderfully sweet and fragrant they are. It's okay to say, "Galias don't grow very well in this area, but I can recommend another variety that is just as delicious."

Next, go to a supermarket with a good produce section and look carefully at the appearance of the vegetables and fruits. Your produce has to look even better than supermarket produce if you hope to sell to an upscale restaurant. It must be cleaned, trimmed, and packed professionally. Presumably, your offerings will also look fresher and be more varied and interesting than supermarket fare.

Remember that chefs are accustomed to buying from wholesalers. They are familiar with the standard box sizes, bunch sizes, and weights used in the wholesale produce business, and they base their recipes and buying decisions on those standards. You need to know what those standards are so that you understand what the chef wants when ordering a box of something or 12 bunches of something else. The USDA grading and packaging standards are printed in the back of this book.

Many specialty produce items, however, are so new that wholesalers don't offer them, so there is no standard size. Whenever you're in doubt, ask the chef what he or she expects. If you're asked for baby leeks, get details about how big a baby leek should be; you don''t want to let them get as big as your thumb if the chef wanted them as big as a pencil. If you're asked for daylilies, take in some buds and some open flowers to find out what will work for the chef. A friendly chef will take you into the cooler and show you what the competition is offering, how it's bunched and how it's packed. Occasionally you will have a market advantage by not offering standard sizes—for example, selling a few bunches rather than a whole case. Be flexible with chefs, and they will value your service as much as your produce.

Once you have established a relationship with a creative chef, he or she may ask you for things you have never considered. One grower reported being asked to grow garlic root hairs for one of New York City's top chefs. Root hairs don't weigh much, so the grower had to negotiate with the chef to be sure he would be adequately compensated for his time. (He found he could grow them by planting small cloves of garlic in trays of vermiculite.) Some chefs toss out ideas like that with no idea how much money and time the grower has to invest, so you have to be careful. One grower who sells to chefs says he smiles and nods when a chef asks for some oddball item, then figures out later whether he can make any money growing it. Many times, by the time the farmer plants, grows, and harvests the crop, the chef has moved on to some other seasonal menu item. Another potential problem when custom-growing unusual items for a chef is the possibility of the chef quitting or getting fired. If you agree to grow something you can't sell elsewhere, be sure the deal is approved by the owner who pays the bills, as well as the chef. One grower recounted growing a half-acre of basil for a chef on a handshake basis. The chef quit in midsummer, and his replacement didn't want the basil. The grower was out of luck.

Even though high-dollar specialty produce may be important to the chef-grower relationship, the big-ticket items for most small growers will be salad mixes, heirloom tomatoes, specialty peppers, fingerling potatoes, baby vegetables, herbs (particularly basil), green garlic (garlic shoots), and, possibly, edible flowers.

❧ HIGH-VOLUME RESTAURANTS ❧

Although specialty produce brings in the most money per pound, some growers can make a good profit by selling a few lower-priced items in volume. Most restaurants, including the high-priced ones, use huge amounts of lettuce, tomatoes, and cucumbers for salads and plate garnishes. If you can supply large quantities of those items throughout the summer, you can win customers at every type of restaurant. You'll be competing with agribusiness and with the cook's buying habits, but you can break into the business if you can improve on quality and service and still keep your prices in the same range as the wholesaler.

Be realistic about pricing. Know your cost of production, and keep on top of current wholesale prices. (More on pricing is in chapter 7.) A distributor adds about 20 percent to wholesale prices; you should mark up the same amount and maybe more if necessary to cover your cost of production and profit margin.

Whatever you do, don't try to win business by undercutting other market growers in your area. You may make sales in the short term, but you will be hurting yourself in the long run. You probably won't be able to make a profit, so you'll get discouraged with your farming venture and possibly go out of business. Chefs won't be happy if you try to raise your prices later, and may just dump you and return to their original supplier. Worse still, they may decide to forget about local growers altogether and revert to buying the cheapest produce they can get on the wholesale market.

A far better strategy for the small grower is to compete by providing higher quality and greater variety. Find out what other farmers are producing for restaurants, and see if you can fill any voids. Talk to chefs about their ideal product—and do your best to supply it. Or try to sell your produce to other restaurants that aren't currently buying locally.

The goal of every new grower should be to expand the market. You don't need to put other growers out of business to be able to start yours. Currently, only a tiny fraction of the produce consumed by restaurants is grown locally. If market growers work together to expand that market, there will be plenty of room for everyone.

► How to Approach Restaurants ◄

Most restaurant owners and chefs will be willing to talk to you about buying your produce if you contact them at the right time. For the grower, the best time to get started is in winter before you order seeds or plan your production. Draw up a list of potential customers, and find out what kind of food they sell and what hours they are open. You don't want to call midmorning if the restaurant does a big lunch business. Try to find out the quietest time in the kitchen, then call to see if the chef would like to sit down with you and talk about what you can grow for them next spring or summer. If the chef doesn't want to even consider the possibility, you can quit right there and move on to the next restaurant on your list.

If you get a nibble of interest, set up a meeting and go prepared with seed catalogs, a list of varieties you have grown in the past, and a willingness to discuss other possibilities. For example, one grower went to a high-end restaurant hoping to sell fresh-cut herbs. The chef, however, said he wanted herbs growing in pots to use in the kitchen, so cooks could snip off what they wanted when they wanted it. The farmer was willing, and a deal was struck.

Most chefs are not going to preorder or sign a contract, though, because they don't know for sure that they will want your produce until they can see it and taste it. Quality is paramount for a chef, so you will have to prove yourself capable. Consider your initial conversation with the chef as brainstorming, not an agreement. It is simply your chance to find out what that chef wants and to decide whether you want to grow it.

Once a relationship has been established, find out how the chef wants to hear from you: by phone, fax, email, or with a standing order. Many growers put together an availability sheet and fax or email it to restaurants once or twice a week; in some cases, the farmer has to follow up to get the order, rather than relying on the chef to call it or email it in, but that's variable from one chef to the next. Find out what the chef prefers, and try to accommodate.

Before your first delivery, ask about how the restaurant pays. Some restaurants will cut a check every time you deliver—and if this is an option, go for it! I know of several growers who never got paid when seemingly prosperous places abruptly closed their doors. If they don't pay on the spot, and most don't, be sure you are very clear about what you have to provide to get paid. Some restaurants pay from your invoice weekly or biweekly; others want a monthly statement that includes all invoices from that month. And if you aren't getting paid according to the procedure you're expecting, ask the chef to clarify. It's sad to say, but when cash is tight some floundering businesses will shortchange their smaller suppliers so they can keep their big suppliers paid.

After quality and variety, dependability is the most important factor chefs look for in local growers. If you can deliver consistent quality, on time as promised, you'll be on your way to success. If something comes up, let chefs know right away so they can get out of a bind. Better yet, always deliver. If you build a relationship of trust, then chefs eventually get less concerned about prices. They know you are being fair, and they will just order the everyday stuff without talking price.

Supermarkets

For most small-scale growers, supermarkets are usually the last place to look for sales. Most market gardeners don't produce in sufficient quantities to meet the demands of high-volume retailers, and the price supermarkets are willing to pay is usually far below the price the farmer can get by direct-to-consumer sales at farmers markets, on-farm stands, and subscription programs.

However, there are some circumstances in which supermarkets can be a good deal for the small grower. Here are a few examples:

* If you don't like to spend time marketing and you're not particularly good at developing a relationship directly with consumers, a supermarket can take virtually all your production off your hands. The question is whether you can make enough money to justify your production time and costs. Supermarkets will pay only half or less of what you could make by direct-marketing at retail prices.
* If you're in a remote rural area and don't want to drive to the city more than once a week to sell your produce, you may find that you can sell everything to a supermarket with just weekly deliveries, provided you have good postharvest facilities, including refrigeration.
* If you can grow just a few crops in large quantities without a lot of extra effort, in addition to your regular diversified garden, supermarket sales can provide a big infusion of cash once or twice a season. Say, for example, that you grow hundreds of things to sell at farmers markets all summer long. You could also plant a few acres of winter squash, which don't require attention until most of your garden is finished in fall, for one big wholesale deal with a supermarket. In general, the best crops for wholesaling are those that don't require a lot of hand labor and those that require the most work at a time when you're least busy with the rest of the garden.
* If you're a certified organic grower and can get a premium for organic produce in a local store, you can make pretty good money. You'll need to find a supermarket with a sophisticated clientele willing to pay extra for organic. In some cases, an upscale supermarket will pay you as much as you can make selling directly to the consumer at a farmers market where people come expecting cheap prices. And those supermarket sales take a lot less effort.

With those caveats in mind, check out the local supermarkets. The produce manager can tell you whether he or she is authorized to buy locally; most chains get all their produce from one central

How Big Is a Bunch?

Many kinds of produce are listed on price reports as "per bunch." To find out what constitutes a bunch, check the standards in the appendix to see if the produce items you're growing list bunch sizes. For example, the grading standards for beets say they are usually sold with 12 beets per bunch with tops attached. In the appendix, you'll see that a box of bunched beets, 24 bunches per case, weighs 36 to 40 pounds. That gives you some guidance about how much each bunch should weigh.

For many produce items, though, there is no official standard bunch size. In those cases, you will have to do a little research. Every time you go to the supermarket, check out produce with an eye toward bunch sizes. Pick up a bunch, observe the circumference at the rubber band, weigh a few bunches, count units if possible. In time, you'll get a sense of what constitutes a bunch of radishes, green onions, spinach, etc. It won't always be exactly the same unit count, and it may not be the exact same weight, but it probably will have a consistent feel and appearance.

warehouse. And usually, the bigger the chain, the more stringent its requirements. Some may require proof that you have product liability insurance (which costs a few hundred dollars a year). Some may have rigid demands about quantities. Nearly all supermarkets will require you to deliver in standard produce packaging. And that packaging will probably have to be new, which means additional costs for you. Supermarkets increasingly are requiring third-party Good Agricultural Practices (GAPs) certification, which can be expensive. For more information on GAPs, see the food safety section in chapter 7.

Natural Foods Stores

The natural foods grocery industry is on a march across America, with several big chains now owning hundreds of supermarket-sized stores. In each of these big, bright stores, the produce department is a major attraction, offering a wide selection of organic, high-quality fruits, vegetables, and herbs.

The proliferation of the natural foods chain stores raises questions for small growers who sell at farmers markets, through subscription programs, and at farm stands.

Will the big stores steal customers from small growers by providing the same quality of organic produce with more convenience? Or will the natural foods stores have a beneficial effect on the market by raising awareness of—and demand for—organic produce in their communities? Or will they at least provide a wholesale outlet where local growers can sell their produce?

To get a better understanding of how market gardeners are affected by the natural foods chain stores, *Growing for Market* talked to growers in communities where those stores have been operating for several years.

For the most part, the growers interviewed did not think that the big stores had hurt their direct retail sales. In fact, many growers noted that the chain stores were eager to buy local produce and had provided them with a new outlet for excess production, which in turn had brought their farms increased visibility. In the case of Whole Foods Markets, which has 340 stores at this writing, the company operates some farmers markets in its own parking lots and has a loan program to help market farmers scale up enough to supply their stores.

However, some cautions emerged from interviews with growers who have been selling to the chains. Several farmers complained that, although the chains talk about supporting local farmers, they will buy only if the locals can meet California prices. Two growers recounted bad experiences in which the stores unexpectedly quit buying in the middle of a crop, leaving the grower with no market. Other growers reported being dropped midseason because the company imposed new food safety restrictions without giving them the opportunity to comply. Most growers, though, said that they get paid on time and are generally well treated by the local store's staff.

In many large cities and university towns, another option exists: the locally owned food cooperative. Most co-ops are retail stores with size and services to rival those of the natural foods chains. Working with co-ops can be a great experience because they operate under a different set of rules: less dollar-oriented and more community-oriented. Even if you don't want to sell to the co-op, the produce manager is a great person to get to know. Often, produce managers will help out local farmers with pricing information, and you'll be amazed at how many other opportunities will arise for working together. If you're just starting out and there is a cooperative natural foods store in your marketing area, get together with the produce manager and have a conversation about your plans. He or she will be a wonderful asset at some point.

Marketing Cooperatives

When a grower can't produce enough of any given crop to justify the costs of marketing it, for whatever reason, a cooperative may be the answer. Perhaps the grower can't produce enough variety to satisfy a CSA; in that case, the grower could work with other growers who each produce different crops for the CSA. In some cases, the grower doesn't produce sufficient quantity to justify a long drive to market; working with others to at least share a driver and truck may be the most efficient solution. Because of the high-volume demands of supermarkets, many growers find they need to work with a co-op to be able to meet the demand. Because chefs require a wide variety of produce items for their menus, they may prefer to work with a single entity rather than having numerous farmers calling and delivering each week.

Packing Tomatoes

If you want to sell to supermarkets, you need to learn packing standards, which are printed in the appendix. Box sizes may seem odd (1⅑ bushel being a standard carton size, for example), and there are other somewhat mysterious standards. If in doubt, ask the produce manager before you pack.

Tomatoes are packed in cartons with the size marked on the outside. Tomato sizes are 4 × 4, 4 × 5, 5 × 5, and 5 × 6. The biggest are 4 × 4, which fill four rows across and four rows down, or 16 tomatoes per layer, in the standard carton. Smaller tomatoes might fill four rows across and five down, or 20 per layer (two layers total).

Other types of tomatoes come in different packs. Cherry and grape tomatoes are packed in 1-pint clamshells, Roma tomatoes are packed loose by weight, and on-the-vine or cluster tomatoes are usually bagged in plastic or mesh bags inside the box.

Maturity stage is also important when selling tomatoes to wholesale markets, and you should consult the buyer to find out the preferred stage. Most long-distance shippers pick tomatoes at the "breaker" stage when they are light green with no pink on them. Local wholesale markets may prefer to receive them when they are pink on the blossom end, because they will ripen in three days at room temperature. For farmers markets, CSAs, and roadside stands, tomatoes should be red-ripe but not soft. For restaurant sales, chefs often want some red and some pink to carry them till the next delivery.

Slicing tomatoes are often packed in layers, such as this box of 4 × 5s. Romas are usually packed loose. Cherry tomatoes are packed in pint clamshell containers or boxes.

What Do Buyers Want?

What do produce buyers look for when deciding where to buy organic and local produce? In a word: professionalism. In several surveys of chefs and supermarket produce buyers, the professionalism of the grower was most often cited as reasons for buying—or not buying—locally.

Reliability and consistency were at the top of the list of criteria cited by 13 retail store buyers and three wholesale buyers in a North Carolina survey quoted in the Carolina Farm Stewardship Association newsletter.

The produce buyers revealed that 50 to 75 percent of their produce purchases were from wholesalers, the rest directly from growers. When deciding where to buy, the produce buyers named these values in order of importance: supplier reliability, attractive appearance of the produce, timing of harvest and delivery relative to their needs, consistent packing and grading, and price or value of the product.

When asked to rank past problems they'd had with local growers, the buyers mentioned inconsistent supply, limited time of availability and limited varieties, difficulty coordinating purchases from many different suppliers, and inconsistent grades and standards.

Buyers said these criteria would be used for selecting suppliers: ability of growers to commit items one week in advance of delivery, consistency in supply quality, and grower experience and familiarity with retail store operations.

In a University of Wisconsin survey, food buyers listed these four top reasons why they don't purchase more local food:

1. Erratic availability; seasonality. Though this was cited as the greatest obstacle, it's obviously not something growers can change but can only educate buyers about.
2. Absence of a central supplier. Buyers complained that no one source could meet all their needs. Seventy percent of the buyers said they would buy more local food if there was "one phone call" they could make to purchase local produce, meat, eggs, etc.
3. Lack of professionalism. Dependability, cleanliness, proper packaging, and product quality were often lacking among local growers, the buyers said.
4. Price. Most buyers said they would buy more if local prices were competitive with the wholesaler's.

Many co-ops made up of small-scale farmers have been successful in the past two decades. Some examples include Vermont Fresh Network, Tuscarora Organic Growers Cooperative, Georgia Grown, and Home-Grown Wisconsin.

Besides the marketing benefits of working cooperatively, farmers also gain from having colleagues to help with problems, offer advice and mentoring, and keep the market served when a member is injured, ill, or called away for family emergency. Many marketing co-ops become buying co-ops as well, allowing farmers to purchase supplies together to get a lower price. When growers share production for a market, they spread the risk of crop failure and extend the season.

Institutional Buyers

Some of the biggest produce buyers have traditionally been overlooked by farmers, but that is now changing. Schools, colleges, hospitals, state park cafes, and other institutions are beginning to buy local produce. A few generalizations can be made about institutions: First, they are usually working within a preestablished budget and don't have a lot of leeway to pay more than they are currently paying to the wholesalers. Second, they don't necessarily have a financial incentive to buy superior-quality produce: Unlike chefs, whose customers will come back if they have a good dining experience, institutional "customers"—students, patients, etc.—don't usually have much choice in the matter. For these reasons, institutions are not likely to be high-dollar customers, but they can buy in large volume, so they are a good match for some farms.

If you would like to explore the possibility of selling to a local institution, first find out how the food services are provided. Many places contract out their food service to a company that specializes in providing meals. If that's the case, you will have to contact the local manager to find out if the company buys from local growers, then probably go into the corporate office to establish a connection. Don't be discouraged—the presence of a corporate food service can actually be positive. For one thing, big companies may already have experience buying local in other locations, so there may be a system already in place. One grower who sells to Bon Appetit, a company that provides food service in many colleges, found that the corporate people were always courteous and respectful, paid on time, and their checks never bounced—all factors that made the food service company a better customer than some restaurants.

If you find interest in buying locally, provide a list of the produce you can supply and when you can supply it. Think big—institutions are not going to buy a case or two of tomatoes for a few weeks in summer. But if you can supply a large volume every week over a long season, you may win a contract.

Food Hubs

You may find that you don't produce enough, over a long enough period of time, to be able to sell to an institution. Don't give up—there may be a food hub nearby that can help. A food hub is defined as "a business or organization that actively manages the aggregation, distribution, and marketing of source-identified food products primarily from local and regional producers to strengthen their ability to satisfy wholesale, retail, and institutional demand." Typically, food hubs are aggregators that buy from numerous small to midsized growers and sell to large-volume buyers, such as supermarkets, colleges, corporate cafeterias, and hospitals.

Food hubs take numerous forms; some are nonprofits, some are privately held businesses, and some are cooperatives. Food hubs include any kind of central coordinator of supply-chain logistics. They provide a wide range of services, such as buying from many small growers to provide sufficient volume to an institution, offering storage and delivery solutions for growers, and processing local produce for sale to large buyers. Businesses that buy from multiple farmers and sell direct to consumers are also considered food hubs. Many food hubs also have developed certification and branding programs to ensure standardization.

The USDA is a strong promoter of food hubs and has many resources about starting, locating, and working with a food hub. Find more information at www.ams.usda.gov/AMSv1.0/foodhubs.

Processors

Traditionally, selling to a processor would be considered a last resort for a vegetable or fruit grower because those markets pay the lowest prices. Today things are changing, at least for organic growers. Organic food companies are paying premiums for certified-organic crops. Bear in mind you have to be a big grower to supply the quantity most processors demand from their suppliers. But if you are a serious organic vegetable or fruit farmer, with significant production

Focusing on Rural Markets

Leah, left, and Marada Cook. PHOTOGRAPH BY LILY PIEL, COURTESY OF CROWN O' MAINE ORGANIC COOPERATIVE

Jim Cook and Kathryn Simonds and their five children started Skylandia Organic Farm in 1994 in far northern Maine. In their first season, they grew about 2.5 acres of vegetables, mostly potatoes.

"At the end of the season, with no local markets and a garage full of potatoes, we unsnapped the seats of our 15-passenger van, loaded a couple of thousand pounds of potatoes, and down the road I went to see who would be interested," Jim wrote.

The Cooks discovered that many people were interested in buying Maine produce—so many that they soon started buying from other local growers to fulfill orders. Thus the Crown O' Maine Organic Cooperative (COMOC) was born. In the years since, the Cook family has developed COMOC into a successful business that picks up agricultural products from as many as 175 Maine food producers and distributes them throughout the state. Jim Cook passed away several years ago, and two of his daughters, Marada and Leah, now manage the business with four other staff members.

Unlike most distributors, who buy from producers in rural areas and sell in cities, COMOC delivers mostly to small towns. "We've worked with buying club coordinators to help develop buying clubs, which is in part an answer to our rural geography, but also the fastest-growing market of our business," said Leah Cook.

They update their availability list weekly and send it to 1,800 recipients. Although most places are on the schedule for weekly delivery, they can increase frequency when the business supports it. They consider flexibility their greatest free resource because they can adjust pickup and delivery schedules to meet demand and build the business.

"We do grower planning every year but have no contracts beyond handshake agreements," Leah said. "We ask our farmers to look at their plans and see where we work for them."

beyond what you can sell locally, you might want to revisit the idea of selling to a processor.

Anthony Boutard, an Oregon grower who sells to Cascadian Farm, owned by General Mills, had this to say: "We have been working with Cascadian Farm for eight seasons, and it has been a productive relationship. They pay well and promptly; they appreciate our quality and never push us to cut corners. They remain dedicated to quality and finding domestic fruits and vegetables. We are tiny producers, but they still deal with us, and the account adds to our economic diversity. I have found it far easier to work with them than the very high-minded local food cooperative. That could change tomorrow, but thus far we have no reason to complain."

CHAPTER 3

The Crops

What should you grow? You probably already have some ideas about what you want to grow, based on your own preferences and your observations about potential markets in the area. That's great—plow ahead. But first read this chapter for information about the possibilities, in case you have overlooked something that could be profitable.

An important fact to keep in mind is that you need a critical mass of product to make marketing worthwhile. It makes no sense to grow only lettuce as an early crop, because there is a limit to how much lettuce you can sell. But if you add spinach, green onions, asparagus, hoophouse strawberries, and peonies to your opening-season menu, you're on your way to a profitable crop mix.

The basic premise of the market gardening model is that you will grow a variety of crops and direct-market them. However, that model leaves a lot of room for experimentation and fine-tuning to produce the greatest return. Some crops are inherently more profitable than others, but you may still need to grow less profitable crops to attract or satisfy customers. Similarly, you may find you make the most money from selling specialty vegetables at farmers markets, but if that doesn't provide enough revenue, you may want to consider growing a few big crops to wholesale.

My point is that there is no single formula for success. You have to grow what yields best, with the least labor, on your farm. You have to grow what sells best at the highest price. You have to grow what fills gaps in your production. And so on. The best advice is to start with the crops you suspect will do best for you, keep good records on expenses and sales, and constantly analyze your results. Most market gardeners tweak their businesses every year—that's part of the fun! Monotony is rarely a factor in market gardening. You will constantly find ways to make your farm better and more profitable.

In this section, you'll find that I give a lot of detail about growing and selling specialty crops, but I mention only a few highlights about the vast majority of common vegetables and fruits. That's because there is ample information available to you about commercial production of most fruits and vegetables, listed in the resources. Specialty produce, however, is not so well documented. Some of the best information about these high-dollar crops has been printed in *Growing for Market* over the years, written by growers all over the country. This chapter compiles that information into a concise overview of specialty produce.

High-Value Crops

Many people get started in market gardening thinking that they are going to build a business on crops that command high prices. Everyone needs to grow some high-value crops, but very few people can make a business out of growing *only* these crops. The fact is that high-priced crops have a limited market, and only a few farms catering to high-end restaurants in big cities are going to be able to subsist entirely on these sales. New farmers also discover that the pricier crops are often a pain to grow, requiring much more labor, more expensive inputs, and better scheduling than more basic crops. So don't count out such basic crops as potatoes and cucumbers when you're starting. But do consider at least some of the following high-value crops.

☛ MICROGREENS ☚

Microgreens are tiny seedlings of salad greens cut at the first true leaf stage of growth. Chefs use them for garnish on a wide range of dishes and add them to salad mixes for extra pizzazz. Growers are reportedly getting very high prices for microgreens; according to one source, chefs pay $25 for a 5 × 6–inch clamshell; according to another, they are paying $75 per pound.

If those numbers make you want to jump into the microgreens business, read on. Microgreens may be extremely profitable for some growers—there are a few large growers who closely guard their production secrets—but they are not necessarily an easy crop.

"What Else Can I Buy?"

John and Karen Pendleton were traditional row-crop farmers near Lawrence, Kansas, when they decided to start a U-pick asparagus business. In 1981, they planted their first half-acre of asparagus. The crop was a success, and they continued to expand their planting, eventually devoting 20 acres to asparagus. Early in their progression from traditional farming to direct-market farming, the Pendletons had a revelation.

"People would come in to pay for $5 worth of asparagus, pull out a $20 bill, and ask, 'What else can I buy?'" Karen said.

So the Pendletons expanded again, this time into different crops and products that could be sold during their asparagus rush. They put up a greenhouse and grew

hydroponic tomatoes. Then they created several value-added products that could be sold year-round, such as pickled asparagus and blue corn chips. Next they added plants, including annuals, perennials, herbs, and vegetables. They grew spinach and rhubarb. They planted several acres of peonies.

In the intervening three decades, the Pendletons have continued to extend their season and add more products. Today they have a farm store that is open from April through November, and they sell at the Lawrence Farmers Market. Cut flowers have become a big part of their business, especially since Karen started doing floral design for weddings and other events. She also dries flowers for crafts and holds several workshops each year for customers to make their own dried-flower wreaths and other home decor items. The Pendletons sell products from neighboring farms, such as shiitake mushrooms, artisan cheeses, and frozen meats. They celebrate the autumn agritourism season by incorporating more pick-your-own crops, a pumpkin patch, family play area, and an educational butterfly house. Most recently, after identifying the need of young families for more convenient access to local produce, the Pendletons developed a CSA program, delivering to local childcare facilities.

By adding crops and ventures wherever they saw a niche, the Pendletons have created a thriving, full-time farm business that has supported their family of five and become an integral part of their community. For more information, visit their website: www.pendletons.com.

Microgreens are usually cut and packed in clamshells for retail sales. For restaurant accounts, they can often by sold by the flat so the chef can cut them as needed.

"It's hard to make money at it," said Tucker Taylor at Woodland Gardens in Athens, Georgia. "It's what draws chefs to our farm, but we'll make our money selling them carrots and tomatoes and salad mix."

Di Cody, formerly a commercial seed sales representative at Johnny's Selected Seeds, recommends that growers start small and do some experimentation to figure out how best to grow them before getting into microgreens in a big way. She suggests starting with six basic varieties that are easy to germinate: Garnet Red amaranth, Bull's Blood beet, red cabbage, Red Russian kale, purple kohlrabi, mizuna, and mustard. She also suggests spending no more than $25 a pound on seed at the start. Any plant with edible leaves can, in theory, be used in a microgreen mix; the trick is to choose varieties that germinate and grow fast, provide a good mix of flavor and color, and have as much weight as possible in the single-leaf stage. Lettuce, for example, is not a good candidate for microgreens because the seedlings are so light and delicate; kohlrabi is a good choice because the seedlings are thick and weighty but still have a pleasant, mild flavor. The Chef's Garden in Ohio, a big producer of microgreens, grows as many as 80 varieties for various mixes, according to an article in the *New York Times*. Lemongrass, buckwheat, basil, fennel, popcorn, and many more have all been mentioned as good ingredients for microgreens mixes.

Microgreens are seeded thinly on a soilless growing medium, such as peat and vermiculite, or on a soil-and-compost mix. The mix should be moistened and spread an inch or two deep in a seedling tray. The deeper the soil, the less often it will need to be watered, but shallow soil depth means lower input costs. Seeds can be either broadcast in flat trays or seeded in lines in channel trays. Density is something you have to learn—you want to get as much production from the space as possible, yet if you seed too thickly, seedlings will stretch and the possibility of damping-off increases.

Most seeds will germinate best at 70°F (21°C). A plastic dome or row cover over the flats will help germination but should be removed immediately after the seeds germinate to prevent damping-off. Seedlings can be misted, but bottom watering is preferred, again for disease control. Big operations use ebb-and-flow watering systems. In general, fertilization is not needed because the seeds contain enough nutrients to get the plants to harvestable stage. The turnaround time from seeding to harvest is 14 to 21 days with most species.

Most growers cut the greens by hand with scissors. That's where the labor gets intensive. At Woodland Gardens on the day we called, 180 trays were ready for harvest.

"We have four people harvesting, and it will take them a good part of a day to harvest, wash, and pack them," Tucker said. "The labor is really expensive."

Eliot Coleman of Four Season Farm in Maine has solved the harvest labor problem. He sells microgreens by the flat, uncut. He has his logo imprinted on attractive wooden flats (the same ones he recommends for soil blocks in his book *The New Organic Grower;* see resources at the end of the book), and he takes the entire flat of seedlings to the restaurants, where chefs cut them as they need them. That idea has caught on, and you will sometimes see micro-greens sold that way at farmers markets.

▶ HEIRLOOM VEGETABLES ◀

Heirlooms are varieties that have been around for decades or longer; because they are open-pollinated (nonhybrid) varieties, seeds can be saved and passed down from generation to generation. For most commercial growers, heirlooms and hybrids both have a place in the market garden. Heirlooms can save on input costs, if you save seeds, and you can select plants that do best in your conditions. Heirlooms are often less expensive seeds than hybrids. And there

may be some marketing benefits of growing heirlooms, thanks to a great deal of information in the media suggesting that heirloom varieties have better flavor. Hybrids can offer many other benefits, such as disease resistance, higher yields, and more uniformity—all important factors to a grower who is trying to make a livelihood.

Heirloom tomatoes have become almost a religion with some growers and eaters in recent years. Most don't ship well, which means they are usually available only from local markets and farms. Their unusual colors, shapes, and histories make for great conversations at market. Heirlooms are often the winning varieties in taste tests.

For many years, growers lamented the disease problems and low yields of heirlooms—until they discovered that grafting heirloom varieties onto hardy rootstocks created much more vigorous plants. Now grafting is becoming almost routine for serious growers, especially those who have had disease problems in the past. Grafting takes practice and patience, but it's not rocket science. If you have a greenhouse, you can probably graft your own tomato plants. If not, you may be able to buy them locally or from a mail-order grower. To learn more, do an Internet search for "grafting for disease resistance in heirloom tomatoes."

Food Politics

I was at a farmers market recently where I saw a young woman approach a display of beautiful, multicolored cherry tomatoes. As she reached out to pick up a pint, she asked, "Are these heirlooms?" The grower replied honestly, "Some of them are." The woman snatched back her hand as though she had been burned and walked away. Obviously, she had the idea that only heirloom varieties were worth buying.

As consumers have become more interested in local food, they've gotten harder to please. They may question growers about all kinds of things a shopper never previously considered, such as where you buy your seeds or whether you use manure in your fields and, if so, what kind of manure. You are likely to be asked if you grow organically; if you use genetically modified organisms (GMOs); if you support Monsanto in any way, shape, or form; if you're GAPs-certified. And that's just the current crop of food issues. Next year, there are likely to be some other topics that have risen to the top of the agenda for informed consumers.

This can be good for you as a grower. It gives you the chance to engage your customers, to have them get to know you better. Listen carefully to what your customers want

A farmer's price sign informs buyers that the carrots are organically grown and heirlooms.

to know, and if you don't always have time at market for a discussion of your practices, express yourself with signs.

➤ BABY VEGETABLES ◄

Baby vegetables are, for the most part, just regular vegetables picked small. Not all vegetables are candidates because not everything is mature enough to eat when it's still small enough to be considered "baby." There's no such thing as a baby tomato or baby watermelon. Veggies that are used as baby vegetables include green beans (although some slender varieties, such as French filet beans, are bred to be small), beets, bok choy, carrots, corn, eggplant, fennel, leeks, summer squash, and turnips.

Baby vegetables are not as popular as they were a decade ago. But some chefs still like them for their delicate flavor and texture. Because baby vegetables are harvested small, they have to be priced much higher than mature vegetables to be able to earn the same price per square foot that you would have earned if they had been left to mature. Before you grow baby vegetables for a chef, find out what size he or she wants, because there can be a lot of different interpretations.

⟫ SALAD MIX OR MESCLUN ⟪

Back in the 1990s, many market gardeners made a good income from the salad mix business. Growers got up to $20 a pound for colorful mixtures of lettuces, greens, herbs, and edible flowers. But when large-scale farmers saw the potential profits, they jumped into

A colorful mix of salad greens and edible flowers for sale at a farmers market.

it, and mechanized and standardized the production. Now every supermarket in America has salad mixes in bags in the produce section. Bulk salad mix is available to restaurants for about $3 a pound.

Still, salad mix can be a profitable crop for some growers. Locally grown greens are usually much higher quality than the shipped-in product, providing better flavor, texture, and a much longer shelf life in the customer's cooler. Many chefs and consumers also prefer the ever-changing ingredients of locally grown mixes. Contamination of bagged spinach and some salad mixes from California with the dangerous *E. coli* 0157:H7 in the summer of 2006 also put a damper on consumer enthusiasm for mass-produced salad mix.

The *E. coli* scare and the obvious potential for foodborne illness from precut produce has resulted in regulation of the salad mix industry—and it may affect even small, local growers. In some states, mesclun and salad mix are now officially considered processed foods and, as such, must be processed in a certified kitchen inspected by the state health department. (See chapter 7 for more on certified processing kitchens.) Furthermore, some buyers require processed food to undergo a stringent process known as a hazard analysis and critical control point (HACCP) plan. An HACCP plan is basically a food safety program that identifies the potential for contamination of food anywhere in the handling process, then sets up steps to prevent such contamination.

Even if you plan to sell salad mix someplace that doesn't require an HACCP plan, you should become acquainted with GAPs—Good Agricultural Practices. GAPs are a set of guidelines that the Food and Drug Administration (FDA) and the USDA have compiled to improve food safety on farms. Every grower should have a food safety program based on GAPs. Some buyers require growers to get certified that they are compliant with GAPs. See chapter 7 for details on food safety certification.

Unless you're doing your washing and packing in a certified kitchen, be careful what you say about your product. Tell customers the mix is *not* ready to eat until after *they* wash it. Many growers sell salad mix in bags, and the bag bears a label saying it must be washed before eating. At farmers markets, growers often sell individual ingredients in separate bins, so customers can create their own salad mixes, again dispelling the idea that the mix is ready to eat. Don't invite regulation of your product and don't risk a lawsuit by someone who assumes it's table-ready like the HACCP-certified product in the supermarket.

How to grow salad mix

Salad mix, unlike microgreens, is generally grown in the soil. It can be grown either in the field or in a hoophouse. Most growers direct-seed with a pinpoint seeder, which makes straight lines that are easy to keep weeded and drip-irrigated. It also can be transplanted from small plugs started in the greenhouse, in the odd case of germination problems when direct-seeding. Although there are plenty of "mesclun mixes" available from seed catalogs, commercial growers usually plant ingredients separately so that an entire planting can be harvested at one time. The exception is with lettuces—if they are similar in days to maturity, the seed can be mixed and planted together because they will be ready for harvest at the same time, and the mixing will already be done in the harvest container. Other ingredients are grown separately to add texture, flavor, color, loft, and (most importantly) weight to the mix. These can include: arugula, mizuna, kale, tatsoi, mustard greens, beet greens, and herbs. The seeds should be planted in the same proportion you want in the final mix. Most growers start seed every week in the amount they know they can sell in one week. That way, the salad mix maintains a size consistency that is appreciated by chefs and other customers. It also helps with scheduling other crops.

A salad mix grower in Maine mechanically harvests beds of greens that were planted in separate blocks by variety. Even small-scale growers find that it's most efficient to grow varieties separately and mix them after harvest.

Small-scale salad mix growers harvest with scissors or by picking individual leaves, allowing the plant to continue growing for a second cutting later. A recent innovation in manual harvesting is the Quick Cut Greens Harvester from Johnny's Selected Seeds. It has cutting knives that are actioned by a cordless drill and a macrame brush to sweep the cut leaves into a collection bag. For larger-scale salad production, European-made mechanical harvesters are available; a 28-inch-wide model sold by Ferrari Tractor (www.ferrari-tractors.com) in California costs about $8,000.

Here's what most salad growers do with the leaves they have just harvested: In an enclosed area, two tubs of clean water are waiting. In the first tub, the water temperature is close to ambient air temperature. The salad leaves are dumped into this tub and swirled around by workers wearing gloves, to remove loose soil. Once immersed, soil particles will sink to the bottom and insects and bits of leaves will float to the top. A net strainer should be run across the top of the water to pick up the debris that floats there. Then the greens are fished out and put into a tub of cold water, which removes field heat, leading to longer shelf life, and helps crisp the leaves.

This is where quality control takes over. The worker must dispose of leaves that have burnt tips, insect bites, or other flaws. Only perfect leaves are used in a salad mix because chefs will be happiest with a mix that is immaculate and attractive.

After inspection and sorting, the leaves can be pulled out of the water gently and placed in woven-plastic baskets to drain. For drying small amounts, some growers just spread the wet leaves on a table made of hardware cloth that serves as a big colander. You have to be careful about how you dry these greens, though; bending or bruising the leaves will diminish the look of the mix and decrease the shelf life of your product.

Many growers use washing machines to spin-dry salad mix. This is best accomplished by putting the washed greens into mesh laundry bags. For food safety reasons, the mesh bags and the washing machine should be sanitized between uses.

Farmers who grow a lot of salad mix usually end up buying a restaurant-sized salad spinner. See the sidebar for information.

After the salad is dried, it goes into a plastic bag inside a box to protect the delicate leaves. Most of the salad mix now produced commercially comes in 3-pound boxes. But local growers have worked out all kinds of sizes and packaging with their restaurant

The Slow Evolution of a Tool

Jonathan Dysinger with the salad mix harvester he invented. PHOTOGRAPH COURTESY OF JONATHAN DYSINGER

Salad mix growers have tried for years to devise a tool that would speed up the harvest process. Eliot Coleman, author of *The New Organic Grower* and *Winter Harvest Handbook*, was instrumental in pushing for the development of a commercially viable tool. Working with Johnny's Selected Seeds, he and several engineers came up with a greens harvester

customers. It should be stored in a cooler at 34°F (1°C), and it will keep for a week or two.

☙ LETTUCE AND OTHER GREENS ❧

If you decide to enter the salad greens market, mesclun is just one of many ways to go. Head lettuce is always in demand, and may be even more so as schools and other institutions increase their purchases of local food. Top-of-the-line restaurants often create their own salad mixes and are interested in raw materials in bulk.

that consisted of a sharp knife attached to a collection basket. The grower swiped the tool across a bed of thickly planted salad greens, and they were pushed into the basket. The tool was introduced by Johnny's in 2009 and won a following among salad mix growers.

It wasn't perfect, though. Growers found that the cut wasn't clean enough to allow regrowth of the salad greens, and that many leaves were lost when they fell under the collection basket. Around that time, John Dysinger, a market farmer in Tennessee, and his 15-year-old son, Jonathan, visited Coleman's Maine farm. "He was interested in tools and asked what tools we hadn't succeeded with yet," Coleman recalled about meeting the teenager. "I showed him our drawings of the so-far-unsuccessful small motorized harvesters and encouraged him to go for it."

The Dysingers returned to Tennessee, where Jonathan started tinkering. "I tried modified hedge trimmers, electric bread knives, and even made Lego prototypes," he said. "It was something that was consuming a lot of my brain energy. I came up with several ideas, but they required running long extension cords and getting shocked every time I pulled the trigger."

Over the next two years, he came up with a prototype that used a cordless drill to power the cutting knives, and a macrame rope brush to sweep cut leaves into the collection basket. He and his dad flew to Maine, where they demonstrated the tool for Coleman and Adam Lemieux at Johnny's, who were impressed. Within a few months, Jonathan had lined up financing and started to manufacture the harvesters.

"The reason his model works and our previous ones failed is because he was smarter than us and realized that it needed more than just sharp blades," Coleman said. "So he came up with that ingenious macrame cord brush, which pushes the greens into the blades and into the basket. Pure genius. Cuts light crops as well as heavy crops. Cuts high as well as low. The kid did good. It's fascinating how often just throwing a new mind at an old problem is all that is necessary. And now he has his own factory to produce them. I am very impressed."

You can negotiate a price by the dozen for baby lettuce and curly endive; by the piece for radicchio and baby bok choy; by the bunch for Kyona mizuna, arugula, and mustard; or by the pound for watercress, garden cress, and mache/corn salad.

❧ Baby and teenage lettuce is usually sold by the dozen, but it sometimes comes in a 3-pound box, which helps to alleviate the problem of variability in head size. Your best bet is to talk with the chef about the size and quantity desired. Don't be afraid to go in and discuss varieties during the winter. Your

Heads of lettuce are displayed in individual boxes at a farmers market in Grand Rapids, Michigan.

need to order seed is a good way to get to know chefs, and let them get to know you. Some chefs prefer lettuce that is about 6 inches tall; growers usually refer to this as adolescent or teenage lettuce.

We grow all our baby lettuces from transplants. Though that seems to be extra work compared to direct-seeding, we avoid thinning, ensure a uniform stand, extend our season by setting out 2-inch plants at the earliest possible date, and save a lot of money on seed. Don't let the head get so big that whole leaves cannot be used in a salad. Pick by slicing the head off at the root joint with a scissors or knife. It is best to pick before 7:00 a.m. to ensure the highest sugar content. Strive for uniformity. We wash the lettuce as described above for salad mix, then place the heads upside-down on a screen table to drip for a few minutes, keeping an eye out for bugs or damaged leaves. We then pack the heads gently into a plastic bag inside a cardboard box. The liner has the same dimensions as the box, but is much taller so that it can be folded over the produce to seal it in once the carton is closed. Once we have packed our boxes,

Radicchio and radishes.

we label them by variety and immediately put them in the cooler at 34°F (1°C). Lettuce treated that way will keep for up to three weeks, and your reputation will grow if customers find that out in their own refrigerators.

❧ Radicchio is often sold by the dozen. Some is still flown in from Italy. Radicchio is bitter-tasting and prized for its color, a brilliant deep red with pure white veins. Not all of the plants head up. Squeeze the heads gently to feel which are becoming solid. Cut the head with a knife below ground level. If you cut too high, the head falls apart. Strip off the outer leaves until you reach those that are beginning to color up. Wash and pack as with lettuce, and they will keep two to three weeks at 32°F to 34°F (0°C to 1°C).

❧ Baby bok choy is a cool-season crop that can also be grown from transplants. The plants are attractive to flea beetles and are best grown under a row cover. Harvest at about 8 inches by cutting at ground level. Trim the end and remove any damaged leaves. The plants are very fragile, so be gentle when washing. Store at 32°F (0°C).

❀ Arugula, Kyona mizuna, and red mustard are packed by the bunch. A bunch is roughly ¾ pound on average, slightly larger than the circle you make with your index finger and thumb. A rubber band or twist-tie is wrapped around the stems and stem ends are trimmed evenly. Arugula leaves are ready to harvest when the plant is 8 inches tall, about six weeks after planting. Always taste them to make sure they have not gotten too hot. The mustards are ready in three to four weeks and should be harvested when they are 6 to 8 inches tall. We have managed to avoid bunching them by packing the greens loosely in regular produce bags and selling by the ounce or in our plastic-lined cartons sold by the pound. The cresses and mache, aka corn salad, often come packed in ice in plastic-lined or waxed 3- or 5-pound boxes. Garden cress can be broadcast and then harvested with scissors. Care must be taken to keep it clean, as a rinse leaves it soggy. Like all greens, it should be immediately cooled. We sow mache (sounds like the "osh" in Oshkosh) in lines. Thinning helps to get a better plant. It grows slowly, so planting in rows makes it easier to weed. Mache likes cooler weather and is a great one to overwinter. It is one of the tastiest and prettiest of all the salad greens. Harvest the entire rosette at the root juncture with scissors. Mache is widely available in clamshells in Europe and is catching on with high-end chefs in the United States.

❧ HERBS ❧

Culinary herbs continue to be high-dollar crops for people who can find an ample market for them. Most herbs are easy to grow and harvest; most annuals start readily from seed, and perennials can be propagated from mother plants kept over the winter.

Herbs can be sold by the pound or ounce to chefs. Or they can be direct-marketed in smaller quantities (at a higher per-ounce price) to consumers. The most popular and easiest to grow herbs are basil, cilantro, dill, fennel, oregano, parsley, rosemary, sage, and thyme. Some chefs may request more unusual herbs such as chervil, culantro, lemongrass, marjoram, mints, papalo, and savory. The best book currently on commercial herb production is *Growing and Selling Fresh-Cut Herbs* by Sandie Shores.

Medicinal herbs are a highly specialized crop with completely different marketing channels, for the most part. There is some opportunity for community herbalists who supply healing herbs and conduct workshops and consultations. A great resource for that line of work is the book *The Herbalist's Way* by Nancy and Michael Phillips.

Growers who are interested in large-scale medicinal herb production will find potential markets among processors, but prices for raw herbs are volatile. Details about growing for the medicinal herb market are available in the book *Medicinal Herbs in the Garden, Farm, and Marketplace* by Lee Sturdivant and Tim Blakley.

Most culinary herbs are sold in flat clamshell containers in supermarkets or in plastic bags at farmers markets. Parsley and cilantro, however, are sold in large bunches.

Lavender Farms

A pick-your-own lavender farm near Sequim, Washington.

Few things are as beautiful as a field of lavender in full bloom—the colors, fragrances, and undulating shape of the rows creates an aesthetic pleasure for growers and visitors alike. No wonder lavender farming is booming in the United States! In a few locations, lavender has become a tourist industry, with farm tours, festivals, and wedding venues during the season of bloom. Many small farms throughout the United States have added a lavender component to their business. The purple herb can be grown for fresh or dried sales, such value-added products as sachets, cosmetics, and food products, including baked goods.

Lavender is not an easy crop, though. It has specific requirements for successful production, including well-drained, alkaline, gravelly soils, plus a sunny, dry climate. It is expensive and time-consuming to establish a field of lavender. It's labor-intensive to harvest. And growers have to develop their own markets for their products.

To learn more about lavender production, visit some lavender farms. Sequim, on the Olympic Peninsula of Washington, has an annual lavender festival and a biennial conference for commercial growers. Blanco, Texas, also hosts a lavender festival each year. Search for lavender farms online and you may even find a farm nearby that you can visit to begin your research.

❧ EDIBLE FLOWERS ❧

This once-trendy item is not as popular as it used to be, but there is always some demand. Chefs may use edible flowers in mesclun mixes or as plate garnishes. Edible flowers are also used to decorate wedding and other special-occasion cakes. Three main cautions about growing edible flowers: Don't grow anything unless you know for sure that it's edible, as many flowers are poisonous; grow them organically, because they won't be washed before use; and inspect them carefully for tiny insects called thrips, which get deep inside flowers and are virtually impossible to get rid of.

Flowers of culinary herbs, such as basil, sage, and thyme, can be used as edible flowers. Other varieties that are commonly grown include nasturtiums, Johnny jump-ups, violets, calendula, anise hyssop, Lemon Gem marigolds, *Hemerocallis* daylilies (*not* Asiatic or Oriental lilies!), and scarlet runner beans.

Edible flowers should be harvested in the morning, after the dew has evaporated and when the flower is fully open. They can be stored between moist paper towels or in a plastic bag in the cooler. Most do fine at 45°F (7°C), but some are better at cooler temperatures. Experiment with varieties to determine which temperature gives the best shelf life.

At retail markets, edible flowers sell for about $3 per 8-ounce clamshell with a mixture of violas and nasturtiums. Ask for the same price at restaurants, too, as you are not likely to make money if you sell them for much less.

Squash blossoms

This is one flower that remains in high demand in upscale restaurants. Squash blossoms should be grown organically, because they won't be washed before use. That usually means they are going to be subject to attack by the ubiquitous cucumber beetle. So the grower of squash blossoms has to get up before sunrise to beat the cucumber beetles. Blossoms should be picked before they are fully

Squash blossoms are widely available in European markets but rarely found in the United States. Growers usually cut the male blossoms and sell them in clamshells.

open, because their delicate petals are likely to get torn in harvest and transportation if they are fully open.

Any kind of squash can be grown for squash blossoms, and if you harvest only some of the male flowers you won't harm squash production. Some varieties produce more blossoms or even all blossoms. At this writing, squash blossom prices range from $20 to $35 per 100.

❧ CUT FLOWERS ❧

Cut flowers are probably the most profitable crop you can grow, provided you can find a good market for them. Successful cut flower growers report gross revenues of up to $35,000 an acre—more than twice what most vegetable growers can earn, and with a lot less labor. Think about it: Vegetable plants send up flowers, then spend the next 30 to 60 days developing their edible parts before they can be harvested and sold. Flowering plants flower, then they get harvested and sold. With flowers, the turnaround time is a lot faster, and there's a lot less time for them to get attacked by insects, disease, or deer, or to get damaged by wind, rain, and hail.

Growers who add cut flowers to their sales often comment with astonishment that people will complain about a $2 bunch of broccoli but not blink an eye at a $10 flower bouquet. From the point of

Flower harvest at the author's farm.

view of the consumer, that bouquet is going to last a week, whereas the broccoli will be gone after one meal. That's why people are willing to pay more for cut flowers; their perceived value is higher. On the other hand, flowers are still considered small luxuries in this country, so they are more susceptible to economic downturns. On the other, other hand, there will always be some demand for flowers: People die, get married, make up after quarrels, and, in general, continue to experience the many emotions and events that are associated with giving and receiving flowers.

In chapter 1, I discussed revenue on our part-time farm. These numbers also give an indication of how much more profitable flowers are than vegetables. To reiterate: In 2001, we grew vegetables and cut flowers on about 4 acres and hired three young people to help us part-time during the peak summer months. Total sales were about $46,000. Expenses were $28,500, including about $8,000 in wages and payroll taxes. That's a net of about 37 percent of income—or $17,000. In 2005, we grew only cut flowers on about 1 acre. Total sales were $32,000. Expenses were about $13,000, including $2,250 in wages and payroll taxes for one employee who worked part-time for two months. Our net income was 59 percent of gross, or $19,000.

In other words, we made more net income from 1 acre of flowers, grown with the help of only one person outside the family, than we made from 4 acres of mixed veggies and flowers, which required three nonfamily workers.

If you are interested in learning more about cut flowers, please see my book *The Flower Farmer: An Organic Grower's Guide to Raising and Selling Cut Flowers*, available from my website, www .growingformarket.com.

➤ OUT-OF-SEASON CROPS ◄

By growing in greenhouses and hoophouses, you can grow crops either earlier or later than their peak season, thus avoiding the low prices that occur when there's a glut of a particular item on the market. Some examples include strawberries in April or May, raspberries in October after frost, and tomatoes just about anytime but midsummer. Of course, out-of-season crops come at a cost, and the cost is usually heating fuel or at least structures, such as unheated high tunnels. It's not possible to make a blanket statement about how much money can be made from these crops, because so many variables affect both production costs and possible selling price. And that can vary even locally from year to year. Take the example of greenhouse tomatoes: A grower in, say, Oklahoma, may be able to produce early tomatoes in a greenhouse with a low fuel cost because there's so much sunshine there. In Massachusetts, though, the same schedule might require much more fuel because of the winter cloud cover. That's why I would advise potential growers to be very skeptical of package deals that include a greenhouse, environmental controls, supplies, plants, and training—a so-called "system" of production. Very little in horticulture can by wrapped up into a neat package, with success guaranteed. My advice would be to carefully study research reports on off-season production and determine what factors would be different on your farm. For example, Marvin Pritts at Cornell University has figured out how to have raspberries producing in a hoophouse after they have frozen in the field. In New York, that's in October, when apples are still being sold and farm markets are still open. That set of circumstances, pertaining both to climate and markets, makes late-fall raspberries a profitable crop. That might not be true on other farms.

Season extension is a huge topic, and I can only touch on it here. Elsewhere in this book I'll talk about growing in hoophouses and under row cover for season extension.

☙ COMMON CROPS ❧

The previous sections were about crops that are considered high value because they command the highest prices. However, many of the more common fruits and vegetables can be highly profitable when grown efficiently and direct-marketed intelligently. A lot depends on the competition at market, as well as the skill of the grower. Tomatoes, lettuce, cucumbers, peppers, carrots—all of these crops can be highly profitable. And even if they aren't extremely profitable in and of themselves, they may open doors to other crops that are. Chefs, for example, always need tomatoes, and if you have high-quality tomatoes at a competitive price, they may also buy high-dollar crops like basil from you. Once you're in the door, you might be able to sell all kinds of things over an extremely long season. Similarly, there are times of year when you need a critical mass of produce to make traveling to market worth the time. So even though salad mix is your big cash cow, you probably won't be able to sell all you could possibly grow. You might want to grow less-profitable crops like parsley, beets, and spinach to sell at the same time. The more variety you have, the more attractive you are to customers.

The key to figuring out all these possible combinations of crops is to keep good records. A farmer once told me that she didn't keep records the first few years she farmed, and she assumed that her lovingly grown heirloom tomatoes, for which she got $4 a pound, were her most profitable crop. Eventually, she started recording her costs, including labor, and her sales. When she analyzed the numbers—surprise!—her most profitable crop was kale. It turns out it was much easier and quicker to grow, so her cost of production was low, and it was quite popular in her upscale market, so her sales were high. Not that this information stopped her from growing heirloom tomatoes—they had other roles to play in her marketing strategy. But at least she was more clear about why she was growing the tomatoes. And she did increase her kale production.

The best way to choose basic crops to grow is to survey your local markets and find out what other people are growing. You'll get a good sense of what grows easily, because it will be available from many growers. You also will be able to identify crops that are overproduced, because prices will be low and supplies overabundant. You might also discover that one grower has already locked up a limited market for an unusual item, such as kohlrabi or celeriac.

With many crops, though, flavor depends on location. For example, you may find that your farm grows the sweetest cantaloupes or the biggest potatoes in the area. Or you may discover that your carrots have a metallic taste. These are all considerations that can only be discovered by trial and tasting.

◈ STORAGE CROPS ◈

Among the most commonplace vegetables are those that can be stored and sold over a longer season. Every farm should have some of these crops. You have more leisure about selling them, and they can ensure that you'll have something to sell in upcoming months even if the weather turns against you. Say you harvest your garlic and onions in July and get them in storage. Then if you have a hailstorm and lose most of your tomatoes the next month, you still have something to take to market. With proper temperature and humidity, some crops can be stored as long as six to nine months. Those that can be stored at least three months include Jerusalem artichokes, beets, Brussels sprouts, cabbage, carrots, cauliflower, celeriac, celery, garlic, horseradish, kohlrabi, leeks, onions, parsnips, dried peppers, potatoes, pumpkins, radishes, rutabaga, salsify, winter squash, sweet potatoes, and turnips.

⚜ GARLIC ⚜

Many market farmers grow a big crop of garlic, which can be sold for months after harvest. Garlic is definitely one of those vegetables that tastes much better fresh from the farm than purchased in the supermarket, which mostly imports softneck varieties from China. It's a popular farmers market item, and there are dozens of named varieties that can be marketed to gain a following. In addition, hardneck varieties have scapes that can be cut off and sold a month or more before the bulbs are harvested. Garlic scapes have a mild garlic flavor and are cut up and used like onions or chives.

Seed garlic should be purchased from reputable growers who get their crops tested for garlic bloat nematode, a devastating pest that is spreading across the United States and Canada. Garlic is

After harvest, garlic should be cured in a warm, dry place with good air circulation. The plants can be bunched and hung to dry, or the bulbs can be trimmed and spread on racks.

planted in fall and mulched after the first hard freeze. It starts growing early in spring and requires little other than weed control until
it's ready to harvest in June or July. Harvested garlic can be hung or
placed on racks in a warm, dry place with good air circulation. Once
the tops die down, the garlic is cured and can be stored and sold for
many months. Most growers, once they get a good crop, save their
own seed to replant. The best guidebook for commercial growers is
Growing Great Garlic by Ron Engeland.

▶ FRUITS ◀

When starting out, most growers think only of the annual crops they
can plant, harvest, and sell within the first season. They consider
tree fruits and brambles as something to graduate into later in their
farming careers. That's fine, but don't wait too long. Perennial fruit
crops can be a wonderful addition to a market garden, but they take
several years to get established.

Some of the easiest fruits to add to your menu are raspberries,
blackberries, and blueberries. And planting a small orchard of

Growing raspberries in a hoophouse can extend the season and improve the marketable
yield of the crop.

apples, pears, or stone fruits like peaches, plums, or cherries, even a dozen trees, will produce large quantities of fruits five or six years from planting.

Before you plant, contact your local Extension for recommendations of varieties and rootstocks. Don't rely on catalog descriptions because there is a great deal of variability in hardiness to both winter's cold and summer's heat. Your state land-grant university has probably done research into the best fruits for your climate. You also will find cultural advice appropriate to your state, such as when to watch for specific pests. Your best source will be veteran orchardists. They can be fonts of knowledge, and some may be at a point in their lives when they are eager to pass on their expertise.

The best way to meet up with these kinds of mentors is through your state fruit growers associations. Ask your Extension agent what horticultural groups exist in your state, and try to attend their farm tours and annual meetings.

▶ BEDDING PLANTS ◀ AND TRANSPLANTS

Market gardeners who grow their own plants often end up producing plants for sale. It happens easily. If you get a reputation for your heirloom tomatoes, say, or numerous pepper varieties, eventually a customer will ask to buy some plants. Growing transplants is a great way to diversify, and you won't lose much if it doesn't catch on, as long as you're already heating a greenhouse for your own transplants.

You'll have to figure out by trial and error what will be in demand in your area. Now that every supermarket and big-box store has a parking lot full of bedding plants in spring, you are going to have to differentiate yourself. Perhaps unusual vegetables will sell. Or cut flower plants, or combination planters and hanging baskets. Sometimes bigger is better; herbs in gallon pots, rather than 4-inch pots, sell well in some markets. Perennials and shrubs are possibilities for direct sales. Succulents are trendy at the moment.

Selling flowering plants retail means either getting them to bloom or showing them in bloom on good signs. Prices should depend on your costs of production. Don't even try to sell as cheaply as the discount stores or you might lose money.

The markets for plants are often the same as for produce: farmers markets and farm stands, certainly, but you might also sell combination planters and live herb plants to restaurants. On-farm

Succulents for sale at the farmers market in Port Townsend, Washington, and flowering plants laid out in neat rows at the farmers market in Traverse City, Michigan.

plant sales are another option for growers who are willing to welcome strangers into their greenhouses.

There are a couple of potential problems with plant sales. If you want to sell at a farmers market, be sure plants are allowed. Some producer-only markets have strict rules about what constitutes a farmer-grown product. If you grow plants from seed, you're probably okay, but if you buy a plug or liner and keep it in your greenhouse for a couple of weeks, you may violate the locally grown rule. Also, in some states, anyone who sells plants at retail must get a nursery license. Check with your state department of agriculture to learn what regulations apply to your situation.

If you are drawn further into the business of selling plants, be sure to subscribe to nursery and greenhouse publications. Some of them are free to the trade. You'll also want to join your state greenhouse growers association and attend one of the many excellent national conferences for greenhouse and garden center businesses.

Eggs and Meat

Although this is a book about raising horticultural crops, I would be remiss if I didn't mention eggs and meat. Many market farmers raise livestock for two reasons: First, animals fit in with farm management plans in that they contribute to soil fertility, help control insects, and graze crop residues. Second, eggs and meat usually can be direct-marketed using the same channels as produce, including farmers markets, CSAs, farm stands, and restaurants. They can be profitable products, and they help many growers carve out a niche in the marketplace that results in sales of other products. Many farmers have no problem selling produce, eggs, and meat simultaneously. A great example is Joel Salatin of Polyface Farms in Virginia. Joel is the author of numerous books, a frequent speaker at conferences, and overall a great model for the farmer who wants to raise animals.

However, selling meat and even, in some cases, eggs, is more complex than selling produce. Each state has its own regulations, and there are federal regulations for meat sold across state lines. Also, some farmers markets won't allow the sale of meat because of food safety concerns. Contact your state department of agriculture for guidance through the regulatory maze.

In summary, the direct marketing of eggs and meat is an evolving issue. Be sure to check the status of all pertinent local, state, and federal regulations before deciding to add livestock to your operation.

CHAPTER 4

Equipment and Tools

The list of products that you could buy for a market farm is virtually limitless. Every single chore that you will do probably has some kind of tool associated with it. You could easily spend $100,000 on equipment, tools, structures, and supplies. You also can start market gardening with almost no capital investment if you're already an avid gardener and have basic gardening tools.

Rather than try to prescribe what you should purchase when starting out, I'll talk about the range of equipment and tools that are widely used in market farming. What you will need will be determined primarily by the scale of your farm. If you plan to start small—for example an acre or less—you can succeed with inexpensive, human-powered tools. If you're going into farming in a big way, you're going to be looking at tractors and implements. Most farms fall someplace in between. The table below, adapted from a workshop presented by the Michael Fields Agricultural Institute, gives you some sense of the range of equipment and tools that are commonly used on farms of various sizes. This is not a prescription, either; it's just a broad-brush picture of useful equipment and tools. Every farm is different.

My advice is to be very conservative about purchasing equipment. Buy only what you need as you grow your business. Preserve your capital until you know for sure what equipment will provide the most benefit.

TABLE 4-1: Estimated Equipment Needs for Various-Scale Market Farms

Size	Seed starting	Tillage	Direct-seeding	Production	Cultivation	Harvesting	Postharvest handling
1–3 acres	Grow lights, hoophouse	Rototiller or walking tractor, custom work	Earthway seeder	Backpack or cart sprayer, irrigation	Wheel hoe, hand hoes, forks, spades	Field knives, buckets, tubs, garden cart	Wash tank, canopy, storage containers
4–6 acres	Heated greenhouse, more hoophouses	35–40-hp tractor	Planet Jr. plate seeder	1-row transplanter	Cultivating tractor	Wagon, boxes, buckets	Roller track conveyor, hand carts, walk-in cooler
7–10 acres	Additional cold frames	40–60-hp tractor, spader, chisel plow	Stanhay precision belt seeder	2-row transplant-er, sprayer	Tool bar implements, e.g., basket weeders	More field crates	Barrel washer, salad spinner, pallet jack

SOURCE: ATTRA publication "Market Gardening: A Start-Up Guide," taken from Michael Fields Agricultural Institute workshop in 2001

Suppliers

Before you start making any purchases, build a library of catalogs and websites from farm and horticulture suppliers. Catalogs themselves are educational, and if you have them on hand you can quickly answer questions about equipment and supplies as they arise. Many horticultural distributors require you to open an account before you can access their online catalogs, so it's wise to do that before you need something quickly. You can start comparison-shopping to find out which company is going to get your business.

At the end of the book, you will find a list of mail-order suppliers whose catalogs you will want to have. You'll also want to get information from regional companies, which often deliver to your farm for free. If they do have to ship via UPS or another service, at least the shipping costs will be lower from a nearby business. To find regional suppliers, do an Internet search for the term "horticultural supplies" and your state. You can also ask at local garden centers to find out about their suppliers. If your Extension service has a local horticulture agent, he or she can be helpful in identifying suppliers.

Many of the tools and supplies used by small-scale farmers are not widely used elsewhere, however. The best way to find them is through advertisements in small-scale farming magazines, such as *Growing for Market*, and on the Suppliers page at www.growingformarket.com.

Tillage

The first thing you have to do before you plant is prepare the soil. The first big question you face is whether you need to purchase equipment for your soil preparation or whether you can hire someone to do it for you. The second question is just how much soil preparation you need to do.

The underlying principle that should be your guide in choosing tillage equipment is to do only what is necessary for the crop you'll be growing. If you are direct-seeding small seeds, you need a fine-textured seedbed, so you will need to do more extensive tillage. If you are planting big transplants, you need less tillage. All tillage has negative effects on soil structure and earthworms, so do only as much as you need. Farmers generally talk about primary tillage and secondary tillage; here are the implements used in each category.

Primary tillage is, as its name suggests, the first and roughest type of soil preparation. It involves turning over crop residues, breaking out pasture, and, when fields have been misused in the past, breaking up deep hardpans that hinder water filtration and root growth. These are the major tillage implements used before planting:

* A *moldboard plow* is probably the most familiar tillage implement. It consists of one or more heavy, curved metal plates attached to a frame; each plate is referred to as a bottom, so a four-bottom plow has four metal plates on the frame. The plow digs deep into the soil, pulling and turning over the top layers, incorporating crop residues.
* A *chisel plow* consists of a series of curved metal shanks that go deeper than the moldboard plow, breaking up the soil layers without turning the soil over.
* A *subsoiler* has straighter shanks than the chisel plow and is used to break up compacted layers deeper in the soil.
* A *disk* consists of a frame with rows (known as gangs) of metal disks that rotate as they are pulled over the soil. They chop and incorporate residues and break up the soil.
* A *rototiller*, or *rotavator*, has powered rotating tines that break up the top layer of soil, providing a fine seedbed.

✤ A *spader* has wider tines that move more slowly, but up and down, breaking up the top layer of soil and incorporating crop residues without turning the soil over. Spaders are thought to be better for the soil because they don't damage soil structure as much as tillers do.

Secondary tillage involves minimally disturbing the top few inches of the soil, primarily for creating a fine seedbed or disturbing tiny weeds. Secondary tillage implements are available in various configurations of spring tines, disks, and shanks. In vegetable farming areas of the country, you may find a huge variety of cultivators, all specific to the conditions in which they should be used. In other parts of the country, you are likely to have much fewer choices and you'll learn to adapt your tillage to your equipment.

For most people just starting in farming, the most sensible course is to hire a neighboring farmer to do most of the primary tillage, such as breaking out pasture or plowing land previously used as cropland. These jobs require a bigger tractor than you will otherwise need, and you only have to do it once per season.

Secondary tillage is another matter. You need to be able to cultivate and turn in crop residues on your time schedule, not a neighbor's. So if you are thinking about buying a tractor, choose one that best fits the chores you can't hire out to someone else.

Tractors and Implements

Most market farmers have at least a walk-behind tractor, also known as a walking tractor or two-wheel tractor. A walk-behind tractor is what many people think of as a garden tiller, but the commercial-scale models actually take many kinds of implements, including a tiller. Most farms also have at least one driving tractor, also called a four-wheel tractor. Some farmers become "tractor collectors" over the years, having several different tractors so they don't have to change the implements too frequently. Most small vegetable farmers start with a 40- to 60-hp utility tractor with a power takeoff (PTO) for implements.

The implements you purchase will be directly related to the scale of your farm and the layout of your fields. The small-scale market garden can rely entirely on human-powered equipment; the large vegetable farm may depend largely on tractor-pulled equipment. Most farms fall somewhere in between. Some of the most common implements used on market farms include:

* *Mulch layers,* which put down and anchor plastic mulch. They can include bed shapers and drip-tape layers, performing three tasks in one pass.
* *Mechanical transplanters,* which are pulled behind tractors while a worker sits on a seat and drops transplants down a chute; the equipment opens the hole for the transplant, then pushes the soil around it and gives it a shot of water.
* *Compost spreaders,* which throw manure or compost onto fields.
* *Bucket loaders,* needed to load compost and useful for moving heavy things around the farm.
* *Mowers,* both rotary and sickle-bar, for cutting cover crops, mowing field margins and paths, etc.

In addition, there are all kinds of crop-specific harvest tools for larger farms. You'll find many excellent publications and catalogs about mechanized vegetable farming equipment. The end-of-book resources give some of them. I suggest you ask local vegetable farmers about their equipment—what they have, and what they wish they had. You'll soon get a sense of what's best for your area and your scale.

The best way to learn how to use tractors and implements is from an experienced farmer, and that is one of the main benefits of apprenticing before you start your own farm. But if you don't have the time or availability to apprentice somewhere else, keep your eyes open for other learning opportunities. Small farm programs will occasionally offer courses in tractor operation and safety; if you have such a program at a nearby college or university, you might suggest that it offer some hands-on training for new farmers.

➤ WHEEL HOE ◄

On the market garden of 5 acres or fewer, one of the most important tools is the wheel hoe. A wheel hoe is an oscillating stirrup hoe blade attached to a frame with a wheel on it. You grab the handles, position the blade in the bed, and walk along behind the hoe. If you attend to your weeds when they are tiny, wheel hoeing is effortless and just like walking. If you wait too long, you'll have to exert more

Choosing to Mechanize

Ryan Romeyn with his Farmall A cultivating tractor and basket weeder.

When Andrea and Ryan Romeyn started farming, they planned to be small-scale growers using hand labor on a few acres. But as they gained experience and a piece of flat land, they started to see the sense in mechanizing and scaling up.

"I learned about the Farmall A and a light went on," Ryan said, referring to an antique tractor with an offset engine that allows for easier cultivation of row crops.

Now, the Romeyns have 20 acres in production at their home farm in Eastport, Michigan, and 4 acres at a nearby location. They grow mixed vegetables and strawberries, have pastured pork, and are beginning to build a herd of Belted Galloway cattle.

"The people of northwest Michigan have to eat, so either we feed them or California feeds them," Ryan said. "If you can mechanize and grow more food without

energy, pushing and lifting the blade. But in either case, the wheel hoe lets you weed about 10 times faster than using a hand hoe.

Some wheel hoe models have interchangeable attachments, such as a four-tine cultivator for preparing fine seedbeds or a furrower for planting potatoes. Blades are also available in several compromising how you take care of the land, I see that as a benefit."

Their farm is certified organic by the Ohio Ecological Farm and Food Association. The equipment has made it productive enough that they can supply 100 CSA vegetable shares and 200 berry shares and sell at four farmers markets per week and to two natural foods co-ops. Just four years into farming their own land, the Romeyns are surprised at how quickly revenue has grown, and they are beginning to feel comfortable that the farm can provide a sufficient livelihood for them and their four children.

Their success is no accident, though. Ryan spent years studying, working at other gardens, and gaining practical skills in a well-planned and pragmatic approach to becoming a farmer. He attended Central Carolina Community College in North Carolina, earning an associate's degree in sustainable agriculture. He took a job doing carpentry so he could learn the skills he would need to build a house and barn. Next, he took a job as farm manager for a nonprofit farm.

Now on their own farm, with house and barn completed, Ryan uses a Kubota 54-hp tractor and an Allis Chalmers G. Tillage that includes a disc, chisel plow, rototiller, and spader. He seeds with Planet Jr. seeders and an Earthway seeder. For transplanting, he has a one-row Holland Transplanter and a two-row with 29-inch spacing. He uses the transplanters on everything but very small plantings and plastic mulched beds.

For cultivation, Ryan uses the Farmall Super A fitted with Buddingh basket weeders, and he switches those out for disc hillers on potatoes. He has multirow cultivators for the Allis Chalmers G. He has a Williams tine weeder and field cultivator for use with the Kubota. He has a boom sprayer for pest control and foliar feeding.

"I also have a straw or old hay chopper for mulching," Ryan said. "I rent a lime spreader from the local farmers co-op for compost applications—it works much better than a manure spreader—and I have a manure spreader. I seed cover crops with a three-point or shoulder broadcast spreader and use a cultipacker to cover the seed, or sometimes I use the disc set shallow. I also use a five-torch walk-behind flame weeder."

"I do feel like I am still learning how to better use this equipment. That is an art in itself, for often the window of use is small for a particular crop or planting, and it's easy to be focusing on another priority while the window is diminishing. I have always been short on equipment-skilled workers, and that's something I'm working on changing."

widths. Wheel hoes cost between $300 and $400, and attachments are about $60 each.

It's no exaggeration to say that, with a wheel hoe, the chore of weeding becomes a pleasant form of exercise. You get the job done without using fuel, breathing fumes, or listening to a noisy engine.

The author's husband, Dan Nagengast, using a wheel hoe.

Dan Nagengast uses a broadfork to prepare the soil for planting in the hoophouse.

Wheel hoeing builds muscles, and if you get going at a good pace, you'll also get a good cardio workout. Who needs a health club?

➤ BROADFORK ◄

Another tool that is a physical pleasure to use is the broadfork, an old Dutch tool that has been resurrected for use on American market gardens. It is a 24-inch-wide fork with two handles and five sharp tines. You grab a handle in each hand, step on the crossbar, and push the tines deep into the soil. Then you rock back slightly, opening and lifting the soil without turning it over. You lift the tool, move it back 6 to 8 inches, and repeat. The effect is to aerate your soil without damaging the soil structure. Broadforking opens spaces for air, water, soil amendments, and microorganisms to move in the soil, improving its fertility, tilth, and biological activity.

➤ HAND TOOLS ◄

Most of the work on a small-scale farm is done by hand: planting, weeding, and harvesting. Even on larger vegetable farms, where tractors do much of the work, hand tools are still essential for some tasks. Buy the best-quality hand tools you can find, and then take good care of them. They will last you for years.

Probably the most important hand tools you'll buy will be hoes. I think of them as falling into two categories: those that you'll use when you're staying on top of your weeds and those that you'll need when you fall behind.

In the first category, you want small, sharp blades that slice small weeds off just below the soil surface while allowing you to get close to plants without damaging them. The aforementioned wheel hoe does a good job of this, but there will be situations where you'll need a long-handled or a short-handled tool. Several types of hoes are preferred by market gardeners, and these are not the big, heavy chopping hoes you find in hardware stores. These are more specialized tools that you will purchase from a specialty farm supply company:

❁ The *stirrup hoe* has a U-shaped blade that oscillates as you push and pull it, cutting weeds in both directions.
❁ The *colinear hoe* was designed by Eliot Coleman with the idea that you shouldn't have to bend or stoop to weed; the narrow,

sharp, rectangular blade is nearly parallel to the soil so you can stand up straight.

❧ Other small-bladed, sharp hoes are available in trapezoid, diamond, and curved shapes. Everyone develops a favorite over time, so choose carefully and try out hoes whenever possible, such as at conferences and garden shows, before buying more than one.

Short-handled tools are also useful for planting and weeding. You'll keep a high-quality trowel forever and use it for planting large plugs. A right-angle trowel is also useful for transplanting, and many people find it to be more comfortable to work with, as it imitates the motion of making a planting hole with your fingers. In the weeding category, a small wire weeder is helpful for tight spaces, and there are several small versions of the long-handled hoe blades. One of my favorite weeding tools is called a Cape Cod weeder, which has a hooked blade that is pulled flat across weedy soil to disrupt small weeds. It also has a pointed end that can be used to hook the crown of a big weed to uproot it. Another favorite is the Cobrahead weeder and cultivator, available with short and long handles.

Finally, a few well-chosen harvest tools will make certain chores a pleasure. A good pocket knife should always be in your pocket, and you should have at least one large harvest knife for cutting broccoli, cauliflower, and other vegetables with thick stems. A pair of high-quality pruning shears (also known as *secateurs*) also will be used for countless jobs around the farm, from harvesting flowers to pruning fruit trees. I have tried every kind of pruner made in my 18 years as a market gardener, and I still value my Felco model 2 pruners above all others. They're expensive, but you can sharpen them often and replace broken parts, and they will last for many years. I also value the inexpensive Felco harvesting shear model 300 for light cutting jobs, such as cutting flowers and herbs, pruning tomato plants, harvesting cucumbers and peppers, etc.

➤ SEEDERS ◄

When direct-seeding more than about 100 feet, it's much more efficient to use a mechanical seeder. There are four basic types of seeders: plate planters, drills, pinpoint seeders, and precision planters. *Plate planters* have a rotating plate that picks up seeds one at a time and drops them into a furrow at a specific spacing. *Drill seeders* are like funnels that drop seed into a furrow; they do not have a spacing adjustment so are used for seeds that are normally planted close together or those that can be thinned. *Pinpoint seeders* provide precision spacing by way of a ground-driven rotating drum. *Precision planters* pick up one seed at a time and space them at intervals as desired. They are the most accurate and, of course, the most expensive.

The most inexpensive brand of seeder is the Earthway garden seeder, which comes with interchangeable plates for different sizes of seeds. The plates are designed to provide the appropriate spacing for the vegetables listed on the plate. However, you can get a wider spacing by putting tape over some of the holes in the seed plate. You can get closer spacing by going over the row twice. The Earthway is a lightweight, push-type seeder and costs about $85 new, although you can often find them on eBay for less. The Earthway is perfectly adequate for a small market garden and is most useful for such larger seeds as beans, melons, beets, and corn.

A more sophisticated seeder is the Cole Planet Jr., which comes with three seed plates with 39 different hole sizes. It is a drill-type seeder that plants seed closely together, and there is no spacing adjustment. The Planet Jr. is a more durable precision seeder than the Earthway, and it is available as both a push type, which plants a single row, or a tractor-pulled seeder, which can be ganged to plant numerous rows at once. Both types cost about $600.

Pinpoint seeders are best used for small seeds that should be planted thickly: lettuce, spinach, and other salad mix ingredients. They require finely prepared soil.

Precision vacuum seeders can cost several thousand dollars per row and can be mounted in gangs to seed multiple rows in one pass.

My advice would be to purchase an Earthway seeder your first year and then wait to decide if a different type of seeder will be worth the money. If you find yourself growing large quantities of a single type of crop, you will probably benefit from having a seeder that works best with that crop. If you plant small quantities of many different kinds of seeds every time you plant, you may decide that adjusting a seeder is more trouble than it's worth. As with all equipment, you have to make the most of your capital, so don't rush to buy things you may find you don't really need.

⚘ SPRAYERS ⚘

A sprayer is an important production tool on every kind of farm. It's needed for insect and disease control, foliar feeding, and deer repellents. It's just as important on an organic farm as a farm where synthetic chemicals are used. Most farm supply catalogs offer a wide range of sprayers. I recently started using one that I think is perfect for the small-scale market garden. Called the Rocket Spray, it is a battery-powered sprayer that rolls on wheels, cart-style, or can be adapted for use as a backpack sprayer. It can be rolled down to the field and has a 15-foot flexible hose that allows us to spray a large swath of a bed without even moving the sprayer. It's powered with a 12-volt rechargeable battery. An on-off switch operates the quiet but powerful pump, which can spray up to 40 feet and up to 80 gallons on a single charge. The price is $200 to $300, depending on the model. For more information, visit www.rocketspray.com or phone Taraco Enterprises at 580-679-3670.

Greenhouses

Everyone wants a greenhouse when they start out because of the promise of warmth and sunshine in the middle of winter. And that certainly is one of the fringe benefits of market gardening. We have always had a small (20 × 30–foot) greenhouse for propagation. When it's not full of plants, we find it useful for other activities: It's a good place for a clothesline and the exercise bike. We have used our greenhouse for kids' pumpkin carving on a rainy Halloween night and for a bentwood trellis workshop in spring. I have several friends who have water gardens in their greenhouses. And one who has a hot tub!

But do you really need a greenhouse? Maybe not. You may be able to find a way around purchasing a greenhouse when you're first starting out if your budget is tight. For example, you could contract with a local greenhouse grower to custom-grow your plants, or you may be able to buy plugs from one of the big plug producers. As a flower grower, I buy many plugs even though I have my own greenhouse, because some varieties need to be started so early (lisianthus is the prime example) that heating the greenhouse in the dead of winter would cost more than purchasing plugs. I know of one growers cooperative that divides up the greenhouse chores among several farms: One starts the greenhouse early and grows everyone's slow-growing plants, another starts a month later, and so on. That saves energy and money for everyone in the group.

Many very small growers also find that a setup of shop lights and shelves is adequate for starting plants in the house. Before you go this route, calculate the amount of space you'll need to start all the plants you want to grow. The key thing to remember is that you want to be able to keep the lights about 2 inches above the plants as they grow, so figure out a way to either move the lights or the shelves at least 6 inches. When we started market gardening, we grew thousands of plants on board-and-block shelves with fluorescent shop lights attached to the shelf above by chains. As the plants grew, we raised the shop lights. We also grew 16,000 transplants one year from a lean-to greenhouse that we made by literally leaning 2 × 4s against a south-facing shed wall and covering them with greenhouse poly. This setup was about 10 feet long by 6 feet wide at the base, big enough to put a heater in when nights got too cold. This was not a permanent solution by any means, but I mention it to illustrate the point that you *can* make do when money is short.

A 30 × 96–foot greenhouse can easily cost more than $30,000 by the time you prepare the site, run water and electric lines, purchase the greenhouse and climate control systems, have it built, furnish it with benches or beds, and supply it with fuel, heat mats, flats, and pots. That's for a hoop-style greenhouse with double poly covering; a greenhouse with rigid covering will cost more.

John Bartok, a greenhouse engineering expert at the University of Connecticut, suggests several ideas that will help you decide how much greenhouse you need, if any, before you start shopping. Here are the issues that will determine how much you have to spend:

- ❀ **Consider your space.** How big does it need to be? You need 2 square feet of bench space for every flat of plants that you need to grow simultaneously. So figure out how many plants you need, based on the size of the cell you use, and go for the closest fit. If you build bigger than you need, you'll be heating unused space. If you expand later, your second greenhouse can be bigger.
- ❀ **Buy local.** Many areas of the country have local greenhouse companies that bend their own pipes and sell greenhouse structures at much lower prices than the national companies. But even if you don't have a small, local greenhouse company, shop first at the national company nearest to you to save on shipping costs and to be sure you get one designed for your area's snow loads and wind conditions.

❀ **Consider your climate.** If you're in an area that gets a lot of wet snow, a peaked or gothic-arch greenhouse that will shed snow better will be a smarter choice than a Quonset or hoop style. If you're going to keep the greenhouse heated all winter, snow will melt and won't be a threat to the structure. However, if you're going to leave it cold most of the winter, choose a stronger structure that can withstand a heavier snow load.

❀ **Get ventilation that suits your schedule.** If you're home on the farm all the time, you can save money by getting roll-up sides and a small fan to keep daytime temperatures from getting too high. But if you're away from the farm all day, you need a ventilation system with a thermostat to take care of cooling while you're gone.

❀ **Consider the alternatives.** Before you buy a big greenhouse that will hold all your plants at once, think about whether you can start plants in a small greenhouse, then move them to an inexpensive cold frame to harden off. A cold frame is much less expensive and can be outfitted with a small heater in case of the occasional spring cold snap.

❀ **Get quotes.** Get two or three quotes to compare equipment, and be sure bids are based on similar assumptions. Good sales reps should have data about snow load and wind velocity in your area. Most small growers find they have to be persistent to get their questions answered. A 1,000-square-foot greenhouse is small potatoes for many greenhouse companies, which sometimes sell greenhouses by the acre.

❀ **Buy a used greenhouse, if you've got less money than time.** A surprising number of greenhouses show up in the classified ads. Normally you'll have to dismantle it, move it, and then erect it yourself. If you do it, be sure to mark every part so you can put it back together correctly. We did this ourselves once when cash was short. We saw an ad in the classifieds for an orchid hobbyist's 20 × 30–foot greenhouse. It was a high-quality structure with rigid sidewalls and endwalls and an inflated poly roof. It had heaters and cooling pads and benches. We paid $1,200 for it, and it's still here nearly 20 years later.

⇝ HOOPHOUSE/HIGH TUNNEL ⇜

In the past decade, the biggest innovation in market gardening has been growing in hoophouses, also known as high tunnels, cold

How Many Transplants Can You Fit?

A standard flat (also called a tray) for starting plants measures about 10 × 20 inches. Flats can be purchased with or without drainage holes. Inserts that fit into the trays come in many shapes and cell sizes. For germinating seeds, we like the inserts with shallow channels so we can separate varieties. Then we transplant into inserts with either round or square cells. These are identified by the number of cells per tray; a good, all-around size for vegetable transplants is 98 cells. Some small plants, such as lettuce, can be grown in 200-cell trays or even 288s. Other plants, such as tomatoes and peppers, fare better in the field when they are bigger, such as those grown in trays with 72 cells.

We rarely find any advantage to growing in bigger cells than 72s.

You'll have to find the best cell size for your own situation. Obviously, the smaller the plugs, the more you can fit in the greenhouse (or under lights) and the less money you'll spend on structure and on heating fuel. Your field conditions will determine how small you can go. If your soil is very fine-textured and you're able to keep it moist while plants are getting established, you can use smaller plug sizes. If your soil has a lot of clumps, or if wind or heat make it hard to keep the top few inches moist, you should go with bigger transplants. For more on growing transplants, see chapter 6.

The unheated hoophouse has become an important tool on most market farms, as it allows growers to extend the season and offer higher-quality products. This one is at Dripping Springs Garden in Arkansas.

frames, or unheated greenhouses. Although there are plenty of variations among the structures that go by these names, the common thread is that they are erected right in the field and plants are grown in the soil inside them. Most are passive solar—that is, they don't have heating systems but rely on sunshine for warmth. In many cases, they also don't have electricity to power fans and shutters but instead rely on roll-up sidewalls or manual vents in the endwalls for ventilation. A 20 × 96–foot hoophouse can be purchased and built for about $3,000 to $5,000.

Hoophouses have proven their worth on virtually every farm where they have been used. They allow for a longer season, as they protect plants from early and late frosts. They permit winter production, because a low inner tunnel of wire hoops and row cover adds enough protection for many cold-tolerant crops to grow in all but the coldest temperatures. They increase quality and yield in most crops. They permit production of crops that would otherwise be impossible because of conditions outside. In all, they give growers a measure of economic security that has never been possible when growing outside.

And though most market gardeners think of hoophouses as something you graduate into with experience, I would argue that a hoophouse is a great place to *start* market gardening. Why? For all the reasons cited above: a longer season, more crop choices, higher yields, better quality. You'll make more money in a hoophouse than in the field—which will speed you on the way to having a viable business. Furthermore, growing in a hoophouse requires less equipment than growing in the field. Once you get the soil prepared, you can handle planting and weeding with hand tools. Growing in a hoophouse is like having a big family garden, only a lot more productive. Another benefit: By dropping plastic curtains, you can zone one part of your hoophouse for heating and do your plant propagation there, thus delaying the need for the much more expensive heated greenhouse.

I definitely would put a hoophouse near the top of the list of tools you should purchase your first year in farming.

How to choose a hoophouse

As you begin to look at cold frames or high tunnels in greenhouse catalogs, you will find a bewildering array of options to consider. Here are the factors that influence price:

* **Size.** With an unheated hoophouse, the best size is the biggest one you can afford that fits in the space available. Taller

greenhouses hold heat better. Sidewalls will cost more, but they allow you to grow right up to the edges. Four-foot sidewalls are nice, and 5-foot sidewalls are even better, so you have the flexibility to grow tall plants on the sides of the house. Although you may grow plenty of small plants—lettuce, for example—that would fit along the sides in a hoophouse without sidewalls, remember that you will need to rotate crops inside the house and eventually you'll wish you could grow tomatoes or cucumbers on the sides and lettuce down the middle.

❀ **Ventilation.** Roll-up or drop-down sidewalls are great for farmers who are around during the day. If you're not around, you can get electric sidewall curtains that open and close according to preset temperatures. Greenhouse experts say that natural ventilation works best in houses that are no wider than 20 feet.

If you don't have venting sidewalls, you will need vents on the endwalls and fans to move the air. The most efficient ratio of length to width is 3:1 if you rely on ventilation from the endwalls.

❀ **Heat.** Although the idea of the hoophouse is to raise crops without heating bills, you may want a backup in case of unseasonable cold snaps. A propane heater will work but should be vented in a very tight hoophouse; growers have lost crops when an unvented heater used up all the available oxygen and the flame went out.

If the sun is shining, a hoophouse can get extremely hot even when outside temperatures are below freezing. That's why ventilation is essential. Drop-down sidewalls, such as those on this hoophouse, are good in cold climates because there's a strip of poly that protects plants at ground level.

Plants can also be covered by row cover held up on wire hoops to prevent the fabric from touching the foliage. Experienced growers have reported 20 to 30 degrees difference between outdoor temperatures and the temperatures underneath row-covered tunnels inside the hoophouse. Eliot Coleman estimates that growing under an inner tunnel inside an unheated hoophouse is the equivalent of growing two zones further south. So if you are in, say, Zone 5, you could be farming like someone in balmy Zone 7.

- **Strength.** Structures need to be strong enough to resist three forces: wind, snow, and weight inside the hoophouse (for example, the weight of hanging baskets or trellised tomatoes). Strength is the major factor separating cheap from expensive high tunnels. Prices vary widely for the same size structure, but so does the amount of steel. Three main components determine the strength of the structure: the *posts,* which are uprights sunk into the ground; the *bows* or *arches,* the curved pipes that make up the roof; and the *purlins,* which are pipes that run the length of the roof. When looking at greenhouse specifications, you'll see numbers like 1.66 O.D. That means the piece is made of steel pipe with an outside diameter of 1.66 inches. The wider the unit, the bigger the O.D. should be.

  You'll also be given a choice in the number of purlins. Those sold as cold frames usually have just one purlin, or ridge pole, running down the center of the frame. Those will work fine in warm areas, if the structure is no more than 20 feet wide. In areas with snow or wind or for wider structures, you may want to go with three or even five purlins. However, if you have sidewalls, the hip boards add stability and may reduce the need for additional purlins.

  One other factor to consider is the bow spacing. Greenhouse manufacturers offer either 4-foot or 6-foot spacing of the bows. Again, if you get a lot of snow, go with closer spacing. If you aren't sure which to choose, visit local nurseries and greenhouses to see which spacing and purlin number is most common in your area.

  Another choice is whether to leave the plastic on the greenhouse during the winter. In areas with a lot of snow, many growers just remove the plastic and leave it off until spring.

  Finally, keeping the structure anchored to the ground must be considered. In the Northeast, greenhouse manufacturers

Haygrove tunnels offer one of the best values in cost per square foot. However, they are designed to be three-season structures, and the poly should be removed in areas where it snows in winter.

say that driving posts 2 feet into the ground is sufficient to keep the tunnel anchored. In the Midwest, though, posts have pulled right out of the ground in high winds, especially if drought has also caused the soil to shrink away from the posts. As a result, many growers in windy areas will pour concrete footings, at least for the corner posts.

⚘ **Endwalls.** Many options exist for the ends of the hoophouse. They can be made of plywood, rigid greenhouse coverings such as Lexan, or studs covered with greenhouse poly. Double doors are great because they allow carts and tractors to enter the house and increase the amount of ventilation. Many growers remove the endwalls altogether in the summer.

Less expensive options

A manufactured high tunnel is a big expense, and if you can't afford one right away, don't worry, because there are other ways to get some of the benefits of a high tunnel at a fraction of the cost.

If you're handy, you can buy, rent, or make a pipe bender to make your own steel hoops from the top rail of a chain-link fence. Johnny's Selected Seeds has developed a hoop bender and offers a comprehensive free manual about how to use it to make what it calls a Quick Hoops high tunnel. Johnny's also sells a smaller bender for making low tunnels from inexpensive electrical conduit.

Many other plans are available for inexpensive season-extension structures. Sometimes they're called "caterpillar tunnels" because the ropes that are strung across the top of the plastic give them a segmented appearance like a caterpillar. You can find more information at www.hightunnels.org and other websites listed at the end of the book.

Movable tunnels

Some high tunnels are designed to be moved, on wheels or rails, so they can be pushed back and forth to cover and uncover crops.

Michael Ableman uses a caterpillar tunnel to extend his season at Foxglove Farm in British Columbia. The name "caterpillar tunnel" is derived from the segmented appearance caused by the ropes that hold the plastic taut across the hoops. PHOTOGRAPH COURTESY OF JOSH VOLK

Movable tunnels give much greater flexibility with crop scheduling and rotations. For example, carrots are seeded in the tunnel in February. By mid-April, they are mature enough to be outside under row cover, so the tunnel is pushed away to open ground. Tomatoes are then planted in the tunnel, where they bear much earlier than field tomatoes. In late summer, spinach is planted outside, where it grows until November, when the tunnel is pushed back over it, extending harvest into December. As the spinach finishes up, the tunnel can be moved again to another crop, such as leeks, which can be harvested until February, when it's time to start carrots again.

Movable tunnels are available as manufactured kits, or they can be farm-built. Johnny's Selected Seeds has a free manual about building a movable caterpillar tunnel.

Coolers

Coolers are a great help on a market garden, because they allow you to spread out your harvest work, rather than picking everything right before you sell it. They also improve the quality and shelf life of your produce during summer's heat. Walk-in coolers are a major expense when purchased new. But they often can be found at restaurant or grocery store auctions for a fraction of the price of a new one. Be aware, though, that moving and reconnecting the cooler will be a significant expense. You also should be sure that the cooler you are considering purchasing can be reused; some older models used a type of coolant that is no longer legal, and the new coolants won't work in them. You'll get good advice if you talk with the people who will do the installation before you buy. Look in the Yellow Pages for heating and cooling contractors, and find out which companies service coolers at local stores or other farms.

Many growers buy refrigerated trucks to use as produce coolers. These can be expensive when the truck is still operable, but once the engine goes, the box and compressor can often be obtained at a low price.

An inexpensive alternative to a commercial cooler is the Cool Bot, a device that modifies a regular room air conditioner to cool down to near-freezing temperatures. It was invented by a small farmer in New York, and thousands of them have been installed on farms in the past decade. The air conditioner with Cool Bot should be installed in a highly insulated space. You'll find advice about building the cool room, as well as ordering information for the Cool Bot, at www.storeitcold.com.

CHAPTER 5

Planning Your Production

One of the most challenging jobs you'll do as a market gardener is to plan and schedule your production. You have to consider a multitude of factors, such as the best varieties, days to maturity of each variety, the amount of each item you think you can sell in any given week, other crops that you should be selling at the same time, and much more. The key to all this will be your planting calendar, which will help you choreograph this intricate dance.

The first things to consider are what you will grow and where you will sell them. Once you have a list of crops that you enjoy growing and that do well in your climate, your next step is to determine how much of each to grow.

The most common mistake new growers make in market gardening is not growing enough. There is a tendency to scale up production from the family garden without any concrete idea of how much that will yield or how much money it will be worth. Too often, the result is that income is far below expectations. A more logical approach is to set a revenue goal and then work backward to determine how much you need to plant in order to generate enough sales to meet your goal.

For the sake of simplicity, I find it helpful to think in increments of $10,000. How much do you need to grow to net $10,000? Let's assume you will be working on a small scale and that you won't hire much, if any, labor, so you can probably expect to net (or keep) about half of the amount you sell. So for each $10,000 you hope to earn, you need to sell $20,000 worth of produce. What's it going to take to get that amount?

First, draw up a list of plants that you want to grow and put the pencil to them each in turn. Let's say you plan to start the season with lettuce. You assume you can sell 200 heads of lettuce at the farmers market every Saturday and 100 heads wholesale to a local grocer each week. You know you have about four weeks of lettuce production in spring before the weather gets too hot. That means you can sell 800 heads of lettuce at the farmers market for $3 each and 400 to a restaurant for $2 each. That's gross sales of $2,400 at market and $800 from the grocery store, for a total of $3,200. So you figure 1,200 heads of lettuce in on your planting plan.

So far, so good. Six more crops like that and you will hit your income target. Next, you turn your attention to tomatoes—how much can you earn there? You may pick a random number like 100 plants, just because it sounds like so much more than you have grown in your family garden. You need to know what yield to expect from 100 plants.

Predicting Yields

This is where planning gets a bit dicey. Yield can vary considerably—by region, by variety, and depending on the weather. Still, you need to find a ballpark or average figure for planning purposes. The best place to get yield information on vegetables and fruits may be your state Extension service. If your state has any kind of vegetable research station, try to get reports of trials. In most states where vegetables are grown commercially, Extension researchers will have grown numerous varieties of important crops like tomatoes and recorded information on yield. Most of these trial reports are now on the Internet, but you might also check with your Extension agent for printed publications. If you can't find anything for your state, look at other states in your region. By perusing these trial reports, you can find high-yielding varieties. These varieties may become part of your crop plan, and you may get some yield figures to plug into your calculations.

❧ CROP ENTERPRISE BUDGETS ❧

Another tool you may find useful in planning your crops are the vegetable crop enterprise budgets that have been done by many state Extension services. Websites are constantly changing at universities, so you may have to hunt around a bit using "crop enterprise budget" or "vegetable crop budget" to see what's currently available. Some of the best examples I've seen have come from North Carolina State, University of Kentucky, and Penn State. Another resource is Vern Grubinger's book *Sustainable Vegetable Production from Start-Up to Market*, which has enterprise budgets for some smaller-scale, direct-market farms. See the resource list for ordering information.

Crop enterprise budgets spell out the costs of growing a specific crop and the returns you can expect from that crop given good, average, or low yields and sold at various prices. Most of these are designed for larger-scale vegetable farms, not intensive market gardens, so their numbers are calculated per acre of each crop. But they are still helpful to give you a sense of how difficult or expensive a crop is and how profitable it can be when grown well.

However, you need to look carefully at these budgets before drawing any conclusions about a crop. Besides being calculated per acre, they may assume mechanical cultivation, in which plants are spaced farther apart than they would be in a hand-cultivated market garden. They also may include routine pesticide applications, which you won't be doing if you are scouting for pests and using biorational pest controls only when needed. Many other variables will be different on your farm. A good crop enterprise budget should explain the production parameters.

Also, check the prices on which the return-per-acre is based. Many of these use wholesale prices rather than the retail prices you will be getting at the farmers market.

Crop enterprise budgets are not a guideline to growing a specific crop in a market garden setting. But they do give you some clues about what to expect. And they give you a format that you can use for your own record-keeping. Get acquainted now with how enterprise budgets are done, because you will want to create some of your own in the future.

An example of a crop budget for tomatoes grown in an unheated hoophouse is reprinted here with permission from University of Missouri Extension publication M170, *High Tunnel Tomato Production*. The complete publication, and a companion publication on high-tunnel melons, is available from the University of Missouri—see the resources for ordering information.

When you peruse this budget, you'll notice that some of the expenses are going to be different from expenses on your own farm. For example, you may pay more or less than $10 per hour labor. (Don't forget to add the employer share of Medicare and Social

Security taxes and whatever workers' comp your state requires.) You may need more or less fertilizer and lime, depending on your own soil test results. Another important point to remember about crop enterprise budgets is that expenses are prorated when possible. For example, you may have to buy 500 tomato boxes at a time, but you'll only use 85 for this crop, so you will attribute only part of your expense to this budget. Look at the fixed costs table and you'll notice that your high tunnel, which costs $3,000, is expected to be used for 10 years, so the cost for a tomato crop is only $300 per year. Although this is perfectly normal from an accounting standpoint, it does not reflect the cash flow issues you'll face as a new farmer. You are going to have to pay up front for most of the materials you use, so your net income is going to be less in the beginning. Later, on the other hand, your profits will seem to be even better than the budget suggests, because you won't have any cash outlay for many of the supplies included in the budget.

Writing the Crop Plan

Once you have located some ballpark yield data, either from trial reports or crop enterprise budgets, you can proceed with planning how much to plant. Let's say you've checked at the local farmers market and tomatoes are selling for $3 a pound. You do some research and find lots of variation in average yield projections, from 5 pounds per plant for field-grown tomatoes to 25 pounds per plant for high-tunnel tomatoes. Maybe after reading several research reports on the subject, you settle on 8 pounds per plant as a conservatively realistic estimate for field-grown tomatoes in your state. That would mean if you sell all you harvest at $3 per pound, you would earn $24 per plant, so $2,400 per 100 plants. But you want to sell $4,000 worth of tomatoes because that's going to be one of your most important crops. So go back and figure out how many plants you need to sell $4,000 worth of tomatoes—166 plants. Then consider that you can't sell an unlimited amount of tomatoes in any given week; you need to spread your production over the longest possible season and offer some choices in types of tomatoes. So you go back to the catalogs and to the Extension reports, and you choose several types of tomatoes, including early tomatoes, main-season

CSA Planting Plan

One of the most admired CSA farms in the United States has made its manuals and planting plans available for free download from its website. Roxbury Farm in Kinderhook, New York, is a community-supported, biodynamic farm that grows vegetables, herbs, and grass-fed pork, beef, and lamb for more than 1,000 shareholders. Its 300 acres of land are permanently protected as farmland by a conservation easement. The farm is operated by Jean-Paul Courtens and Jody Bolluyt.

Among the documents that are available are a seeding and planting plan for 100 CSA shares, greenhouse seeding schedules, a planting manual, a harvest manual, advice about buying equipment, and a soil fertility manual. Together, these documents represent some of the most valuable planning tools I have ever seen, and they are all available for free.

"It is like CSA; the vegetables are for free if you support the farm; the manuals are for free if the outcome is a stronger community and more sharing of information amongst small farmers," Jean-Paul said. "What is important to understand is that they were written for our apprentices to have an insight into how our operation works, what is expected of them, etc. It was never my intention to create a manual that would cover all bases for all kinds of different operations. It is just another way to improve communication on our farm. Other farmers in the Collaborative Regional Alliance for Farmer Training (CRAFT) program (craftfarmapprentice.com) asked if I was willing to share them. So, I put them online. I encourage people to create their own standards and to create a manual for their farm."

With Jean-Paul's caveats in mind, I suggest that his manuals provide an excellent example of the types of information you should be compiling about your own farming systems. In addition, they are a good starting place for beginners who are trying to put together a business plan but have no idea of potential yields and labor requirements. The harvest manual, for instance, has details for 53 kinds of vegetables. For each, it lists yield, value, hours to pick and wash, tools needed, readiness indicators, harvest procedures, washing procedures, and packing standards.

The manuals are updated every other year or so to reflect changes in the farm's practices. The manuals and other Roxbury Farm documents can be downloaded at www.roxburyfarm.com/files.

tomatoes, cherry tomatoes, Roma tomatoes, and heirloom tomatoes. You also can schedule several plantings of your main-season tomatoes so that you have some insurance against a late frost, early blight, tomato hornworms, or any of the other myriad things that can ruin your crop. Many people put considerable energy into having early tomatoes—they grow them on black plastic mulch, cover with row cover, or surround them with Wall-O-Waters or individual hot caps. That's all good (though a hoophouse would be even better), but you might also think about how to extend the season on the back end—by taking good care of your indeterminate varieties so they keep producing and by planting a late succession of tomatoes so you have a fresh crop coming on later in the season.

Spend some time playing with all these calculations, and eventually you will come to a realistic expectation of how much you should plant your first year. That crop plan will become the baseline for all your future planting plans. Every season, you will test the assumptions of your crop plan and make refinements as you have more results to consider. Were you really able to sell 200 heads of lettuce at every farmers market? Could you have sold 300 heads? Did your lettuce season last four weeks? Or did it last five weeks? Once you know the market is there for your lettuce, you can figure out how to have it earlier and later. Every year, you will tweak your plans—to meet demand, to maximize efficiency, and to increase revenue.

Scheduling

After you have completed the exercise of figuring out how much to grow in order to meet income targets, you can get to work filling in your planting calendar. You can do this with a pencil on a big paper calendar or use a computer. First, mark your average first and last frost dates. The field season revolves around these two dates.

Then mark your day length for the first day of each month. For many plants, growth is ruled by photoperiod, which is the ratio of daylight to darkness. So it's useful to become familiar with day length, the number of daylight hours between dawn and dusk. Day length depends on the time of year and your latitude (distance from the Earth's equator). *The Old Farmer's Almanac* always includes tables and charts that allow you to identify your latitude and calculate day length. Or you can find some quick calculators online that will do it for you. Check the list of resources for websites.

Most plants don't grow well with fewer than 11 hours of daylight, even if you can protect them from cold weather, so it's useful to get in the habit of thinking in terms of day length. This will be especially helpful if you plan to extend the season with a hoophouse or grow all winter in a heated greenhouse. Also, onions are bred

HIGH-TUNNEL TOMATO BUDGET

TABLE 5-1: High-Tunnel Tomato Budget per 1,000 Square Feet (170 plants)

Production expense	Unit	Quantity	Price ($)	Labor (rate/hr)	Type*	Hours	Total cost ($)
Soil preparation							
Soil test	Complete tunnel		7.50	10.00	M	0.5	12.50
Major tillage			4.00	10.00	M		9.00
Rototill			2.00	10.00	M		7.00
Compost	ton	1	35.00	10.00	M	1.0	45.00
Raised bed formation				10.00	M	3.0	30.00
Fertilizer and lime	lb	50	5.00	10.00	M	0.5	10.00
Plastic mulch	linear ft	300	5.00	10.00	M	1.5	20.00
Irrigation drip tape	linear ft	300	5.00	10.00	M	0.5	10.00
Plant costs							
Transplants (including seed)		170 plants	26.00	10.00	M	1.5	41.00
Starter solution	lb		1.00				1.00
Production costs							
Herbicide	NA						
Insecticide			6.50	10.00	M	1.0	16.50
Fungicide			7.00	10.00	M	1.0	17.00
Cultivation	NA						
Irrigation/fertigation		1,000 ft³	37.00	10.00	M	5.0	87.00
Twine and pruning				10.00	M	3.0	31.00
Fuel and oil							1.50
Plastic, stake removal				10.00	M	3.0	30.00
Row covers				10.00	M	1.0	10.00
Stakes				10.00	M	2.0	20.00
Wire hoops				10.00	M	1.0	10.00
Temperature management				10.00	M	10.0	100.00
Harvesting costs							
Picking				6.00	H	20.0	120.00
Postharvest costs							
Boxes		85	85.00				85.00
Grading				6.00	H	20.0	120.00
Marketing costs							
Packaging/delivery				10.00		10.0	100.00
Total production costs							933.50
Total fixed costs (from Table 5-2)							155.25
Total costs							$1,088.75

*M = Manager labor; H = Hired labor. See Notes for Table 5-1.

Notes for Table 5-1

Production expense	Comments
Soil preparation	
Soil test	Contact your local university soil-testing laboratory for analysis.
Major tillage	Remove old crop residue (including roots) and till the soil.
Rototill	
Compost	Have compost tested for pH and soluble salts.
Raised bed formation	Make a raised bed 6–10 inches high.
Fertilizer and lime	
Plastic mulch	Use 1–1¼ mil embossed plastic mulch.
Irrigation drip tape	Use 8–10 mil drip tape with 4- to 12-inch drippers. Consider recycling drip tape.
Plant costs	
Seeds	
Transplants	Transplants should be stocky and 5–6 weeks old.
Starter solution	Use a soluble starter fertilizer such as 20-20-20 or 9-45-15.
Production costs	
Herbicide	No herbicides are needed within the high tunnel.
Insecticide	Scout plants near sidewalls and vents for insect invasion.
Fungicide	Scout plants and prevent buildup of humidity.
Cultivation	Very little cultivation is needed during tomato production.
Irrigation/fertigation	Use in-line filters and clean water.
Twine and pruning	In a 20 × 96-foot-high tunnel, approximately 740 feet of twine is used per row of tomatoes.
Plastic, stake removal	Many crops that follow tomatoes (peppers) can be planted in the existing plastic mulch.
Row covers	Row covers are reusable for several years. Keep them clean.
Stakes	
Wire hoops	Wire hoops are $^3/_{16}$ inch × 64–76 inches. Space hoops 2 feet apart.
Temperature management	Monitor temperature carefully. Frequent adjusting of vents and row covers is necessary during flowering in mid-April.
Harvesting costs	
Picking	Two to three harvests per week will be necessary.
Postharvest costs	
Boxes	Use clean boxes.
Grading	
Marketing costs	
Packaging/delivery	

Author's Note: The enterprise budget was created from research conducted in Missouri prior to 2004. As a result, the dollar figures are now out of date and should not be used for specific guidance about the costs involved in high-tunnel production. The types of expenses, however, remain valid and provide a good illustration of all the factors that need to be considered when creating crop budgets for your farm.

TABLE 5-2: Fixed Costs for a 2,000-Square-Foot High Tunnel

Item	Cost	Years used	Yearly costs
Land charge/rent	$100.00	NA	$100.00
High-tunnel material	3,000.00	10	300.00
Plastic covering	300.00	3	100.00
Shade cloth covering	220.00	10	22.00
Wire hoops	40.00	5	8.00
Row cover	20.00	5	4.00
Stakes	10.00	5	2.00
Interest on land and buildings	20.00	NA	20.00
Taxes on land and buildings	10.00	NA	10.00
Depreciation on machinery	10.00	10	10.00
Interest on machinery	10.00	NA	10.00
Depreciation on irrigation equipment	10.00	5	10.00
Interest on irrigation equipment	5.00	NA	5.00
Depreciation on packing building	10.00	10	10.00
Interest on packing building	5.00	NA	5.00
Insurance	5.00	NA	5.00
	Total fixed costs for each year		$621.00
	Area portion for 1,000 square feet		0.50
	Yearly portion used for tomatoes		0.50
Tomatoes total fixed costs			$155.25

TABLE 5-3: Income Sensitivity: Total Revenue ($) per 1,000 Square Feet (170 plants)

Price/lb	Yield per plant (lbs)		
	8	10	12
$1.00	$1,360	$1,700	$2,040
1.10	1,496	1,870	2,244
1.20	1,632	2,040	2,448
1.30	1,768	2,210	2,652
1.40	1,904	2,380	2,856
1.50	2,040	2,550	3,060
2.00	2,720	3,400	4,080
2.50	3,400	4,250	5,100
3.00	4,080	5,100	6,120

Tables 5-1, 5-2, and 5-3 are reprinted with permission from "High Tunnel Tomato Production" by Lewis Jett at University of Missouri Extension. This publication and another on high-tunnel melons are available for $10 each. To order, go to extension.missouri.edu/M170, or contact MU Publications, 2800 Maguire, Columbia, MO 65211; 800-292-0969.

to be grown in certain areas based on day length, so you'll need to understand the interaction between day length and temperature in your location to be able to choose the best onion varieties and grow them successfully.

Once you have your frost-free dates marked, line up your varieties according to when they can be planted outside. Mark those dates on the calendar, too. If you are going to grow the transplants yourself, count backward from the planting date to determine the seeding date. For such heat-loving crops as tomatoes, peppers, and melons, set a later date when the weather is warm—there's no sense in rushing those summer crops because they aren't going to thrive until it heats up anyway. Cool-loving crops, such as peas and spinach, can go a couple of weeks earlier. If you're going to be planting in a hoophouse, you can schedule your first plantings as much as a month earlier than field plantings.

Although some growers still do these calculations on a paper calendar, most now use spreadsheets for planting plans. Johnny's Selected Seeds has a calculator on its website that you can use as a model for creating your own calculations.

❧ SUCCESSION PLANTING ❧

To this point, you've done what every good backyard gardener does: starting in a timely fashion. When you're growing for market, though, you have to keep planting as long as you possibly can hope to get a crop. Some crops (lettuce, for example) may be planted every week, up to the point where the weather will become too hot. Others can be planted every week to 10 days, up to the point where you can't get them harvested before the first frost in fall. This is the more elaborate planning process, and it will require a lot of knowledge about your climate and the crops you are growing.

Season extension is the key to success in market farming. As long as you have a market, you should have as much to sell as possible. Experienced market gardeners work hard at extending the season for every crop they grow. They use row cover to protect against late and early frosts in spring and fall, black plastic mulch to warm the soil for heat-loving crops, shade cloth to keep cool-weather crops going into summer, succession plantings on almost every crop, and careful variety selection to stagger harvests.

Planning for all these crops is probably the most challenging job you will do, but I want to emphasize again: It's the key to your

How Late Can You Plant?

Many varieties of vegetables can be planted in mid- to late summer for fall harvests. Succession plantings of warm-season crops (such as corn and beans) can be harvested up until the first killing frost. Cool-season crops (including kale, turnips, mustard, broccoli, and cabbage) grow well during the cool fall days and can withstand light frosts. In a hoophouse, plants will actively grow until day length decreases to fewer than 11 hours. Timely planting is the key to a successful fall garden.

To determine the time to plant a particular vegetable for the latest harvest in your area, you need to know the average date of the first killing frost and the number of days to maturity for the variety you are growing. Choose the earliest-maturing varieties for late plantings. The formula below for determining the number of days to count back from the first frost will help determine when to start your fall garden.

Number of days from seeding or transplanting outdoors to harvest
+ Average harvest period
+ Fall factor (about two weeks)
+ Frost tender factor (if applicable) (two weeks)
_____

= Days to count back from first frost date

The frost tender factor is added only for those crops that are sensitive to frost (corn, beans, cucumbers, tomatoes, squash, etc.), as these must mature two weeks before frost in order to produce a reasonable harvest. The fall factor takes into account the slow growth that results from cool weather and short days in the fall, and amounts to about two weeks.

Example: Bush beans

    Average first frost—October 15
    Days to harvest—55
    Average harvest period—15 days
    Fall factor—14 days
    Frost tender factor—14 days

Last planting: 98 days before October 15, or July 7.

SOURCE: Virginia Cooperative Extension.
www.ext.vt.edu/pubs/envirohort/426-334/426-334.html

success. In the first few years of farming, you will just have to guess at how long you can extend the season for each crop. As you gain experience, you will refine your planting calendar over and over.

Record-keeping

The only way to benefit from your experience is to keep records of everything you do. You may think that you will remember when you planted which variety and when you started to harvest it. And there may be a few geniuses out there who really can remember all the details on their crops. But when you're growing five, six, seven varieties of 20 different crops, you are not going to remember it all. So how do you keep track?

➤ PRODUCTION RECORDS ◄

Many growers prefer to use crop sheets in a three-ring binder that they keep in the barn or farm office. Crops are alphabetized, with a separate sheet for each variety and each succession planting. Data to be captured include the source of the seed, the amount purchased, seeding date, planting out date, number of plants set out, location in the field, and any special treatment, such as row cover or plastic mulch. Other events in the crop's production (weeding, mulching, or spraying, for example) also should be reported. The crop sheet also is used to record harvest data: date of first and last harvest, amount picked, amount not saleable, amount sold, and price. Really good growers also record the time spent on each task, so they can create enterprise budgets similar to those described above. That way, they have all the data they need to compare crops on an equal footing, which will help them determine which crops make them the most money.

Although many growers still prefer the pencil and paper for this information, which can later be transferred to a computer program, many now use smartphones or tablets to record the information in the field and electronically transfer it to the office computer.

Many growers create their own databases and spreadsheets for keeping records. Others prefer to purchase software programs designed specifically for market farms. Several are listed at the end of the book.

TABLE 5-4: Crop Record

CROP:

Variety:

Seed Source: Amount Purchased:

Seedling date	Transplant date	# Plants	Planting-out date	Location	Comments

Tasks and Time Spent:

Harvest date	Planting/location	Amount picked	Amount saleable	Amount sold/price

Wherever you keep your data, whether it's on paper, your phone, your desktop computer, or the Internet, getting these planting and work records into a database is valuable, because you can then sort and select records in numerous useful ways. For example, say you put your planting successions into a spreadsheet. Once all your varieties are recorded, you can then sort according to seeding date, and you'll have a list of all the things that need to be planted on a given day. Or suppose you are considering buying a new greenhouse and trying to calculate the space you need. You can easily find how many flats of transplants you grew last year if you entered the correct information for individual crops.

❧ SALES RECORDS ❦

Production records are only half the task, though. You also need to keep sales records. Again, you can do it manually, by recording marketable yield when you pick a crop, the price you got for it, and any unsold amounts. Or you can keep these sales records in a software program such as QuickBooks or Quicken. With QuickBooks, the easiest method is to create invoices for every sale, with specific crops in your Accounts list. You also can do the same on a deposit record if you handwrite your invoices and can correlate them with the payment later. In Quicken, a less expensive and easier program, you can do much the same. The key is to create categories for each item you grow, such as Arugula, Beets, Carrots, Dandelion Greens, and so on. Then when you make a deposit, you can specify how much each crop contributed to the total deposit. You also can Create New File, which sets up a file that is not part of your regular checkbook account. Either way, QuickBooks and Quicken can quickly create reports that will help you look at your sales from all angles.

The combination of production records and sales records is essential to improving your profitability in future years. These two sets of records will help you determine whether a crop is really worth growing, as opposed to whether you simply like growing it or think it grows well for you.

CHAPTER 6

Planting and Tending Your Crops

Detailed advice on growing plants is beyond the scope of this book, but there are many systems, tools, and supplies that you need to learn about when you start farming. This chapter will give you a brief overview of horticulture on a commercial scale.

Starting Seeds

Winter is the time for planning, because late winter/early spring is the time for growing. If you are growing your own transplants, you will want to start about six weeks before you will plant them in the hoophouse or field. Here are the items you'll need to start your own seeds:

Greenhouse Trays and Inserts

Channel trays allow for starting a large number of seeds on a heat mat, while using very little germinating mix. The tray in back is covered with an acrylic dome to hold in moisture.

Molded polystyrene trays come in numerous configurations, with square cells, round cells,

and tapered cells in many sizes. They are named somewhat confusingly in catalogs, so here's what you need to know:

* Channel trays are used for germinating seeds. The channels are very shallow, about half the depth of a regular plug tray, so they don't use a lot of seed-starting mix just to get the seeds germinated. They also make it easy to separate and label varieties. And if one variety should suffer damping-off, the channels slow the spread of the disease. Channel trays are only useful for hand transplanting. Once the seedlings have one to two sets of true leaves, we pick them out and transplant into a larger cell tray. The benefit of growing this way is that we don't have any empty cells when we transplant, so we know exactly how many plants are being set out.
* When growing on a larger scale, growers plant seeds directly into a plug tray using a vacuum seeder, which plants one seed per cell. Here's how it works: A seeder

Seeds can be sown thickly in channel trays. As soon as the first set of true leaves emerges, the seedlings are removed in clumps from the channel, roots teased apart gently, and each seedling planted into a cell in a plug tray. It may seem labor-intensive, but it goes really quickly with experience.

tray has holes that line up with the cells in the planting tray. When a vacuum is turned on, the suction pulls a seed to each hole and holds it there. Excess seeds are poured off. When the suction is released, the seed drops into the planting tray. Vacuum seeders start at about $650, but some growers have made their own. For an example of a homemade vacuum seeder, visit www.hightunnels.org.

❖ The standard-sized tray that holds the inserts measures about 10.5 × 21 inches and is referred to as a 1020 tray.

Inserts are the containers in which plants are grown; they come as a sheet that fits into the 1020 tray but can be broken apart into smaller packs. They are used mostly for plants you want to sell rather than those you are growing for your own fields. Inserts are identified by the number of cells in each pack and the number of packs in each flat that fit in the 1020 tray. A 606 insert has six packs per flat, with six cells per pack (the "six-pack"

of bedding plant nurseries) for a total of 36 plants per tray. A 1004 has 10 packs per flat, each with four cells. An 1801 has 18 individual packs per flat.

❖ Propagation, or plug, trays have various numbers of cells per flat, and they can't be broken apart into smaller packs. They are named simply by the number of cells in each tray and the shape of the cells: a 128 square or 98 round, for example. These are the best choice for growing your own transplants. When you purchase plugs from a plug supplier, you can usually choose from a range of sizes, everything from tiny plugs, 500 per tray, to big plugs that fit 72 or even 36 per tray. We find that 210s are about as small as we like to transplant into the hoophouse and 98s the smallest size we transplant into the field. Often we'll order small plugs, about 400 per tray, and bump them up into either 210- or 98-cell trays, depending on where they're going to be planted. We like to grow our tomato transplants in 72-cell trays.

A 72-cell tray is good for large plants, such as tomatoes, peppers, and melons. A 128-cell tray is for lettuce and other small or fast-growing seedlings.

- **Water.** You must have access to clean water in your greenhouse. Ideally, you will be able to heat the water before you deliver it to plants. Dousing tender young seedlings with water that is barely above freezing is not good for them.
- **Heat.** First, figure out a way to provide bottom heat to seeds, to help them germinate. The safest and most reliable way to provide bottom heat is with an electric heat mat with a thermostat to control the temperature. Propagation mats are widely available from greenhouse supply companies and are highly recommended as a good place to spend your money. Many growers have made do with other systems, including hot compost in a cold frame, a water bed, an old refrigerator with shelves and heat cables, and various other ingenious devices. If you can provide 70°F (21°C) bottom heat, your seeds will germinate quickly and speed up the process.
- **Containers and trays.** The cheapest and most readily available containers are molded black polystyrene. If you handle them carefully and store them after use away from sunlight and excessive heat, they will last many years.

Some newer cell trays are made of rigid plastic and don't require a carrying tray, but most still need a bottom tray or flat. Get trays with drainage holes if you have a greenhouse, without holes if you're growing under lights in your house. Get several types of inserts—channel trays for germinating seeds and a couple of cell sizes for growing your transplants. We like to have inserts with about 200 cells for lettuce and other quick-start crops; 72 cells for big crops, such as melons, tomatoes, and peppers; and 98 cells for everything else. Over time, you'll figure out what size transplants work best in your soil and climate, and you may eventually switch to different sizes. But these recommendations will get you started.

Another option is the Speedling tray, which is made of Styrofoam and provides a deep, inverted pyramid cell. These have the advantage of directing root growth downward, avoiding the problem of circling roots, and the tapered shape allows the plug to be removed easily. The disadvantage is initial cost—about double the price of polystyrene—and the fact that they don't nest, so take up more storage space. However, Speedling trays also will last a long time if cared for properly. These are the type of trays known as float trays by tobacco growers. If you're in an area where tobacco has traditionally been grown, you might find many of these on the resale market as tobacco farmers decrease production.

❧ SOIL BLOCKS ❧

Besides trays there are soil blocks, preferred by many experienced organic growers who don't want the expense or the waste of plastic trays, inserts, and pots. Soil blockers compress moist soil mix lightly so that it hangs together on its own at first, then with the help of roots as the seed germinates and grows. The air space between soil blocks becomes an invisible wall to plant roots. Rather than circling, as roots do when grown in plastic pots, roots in a soil block fill in the

Using a handheld soil blocker, a grower makes small blocks for tiny seeds, such as lettuce.
PHOTOGRAPH COURTESY OF JOHNNY'S SELECTED SEEDS, WWW.JOHNNYSEEDS.COM

Lettuce seedlings are ready to plant when the roots hold the soil block together. Photograph courtesy of Johnny's Selected Seeds, www.johnnyseeds.com

block, then stop growing when they reach the air space between blocks. As a result, those roots are ready to grow immediately when they are transplanted into the soil.

Soil blockers are metal molds with plungers that eject the formed soil blocks. They are available in several sizes and styles. They can be outfitted with various pin sizes for making holes for seeds or smaller soil blocks.

The smallest and least expensive are the handheld blockers. They are available as a single 4-inch block for the largest transplants; as four blocks about 1¾ inches square; or as 20 miniblocks, each about ¾-inch square. Miniblocks are used for the smallest plants, such as lettuce, or for germinating seeds of plants that will be later transplanted into the larger soil blocks.

A step up in sizes, prices, and efficiency are the standup soil blockers. One model makes 20 blocks, each 1½ inches. The other makes 12 blocks, each 2 inches. Phone Johnny's commercial department at 877-564-6697 or visit the website www.johnnyseeds.com.

For larger-scale transplant production, the Dutch have developed automatic soil blockers. The Dutch used-equipment dealer Duijndam Machines sells Flier soil blockers for about $1,500. Customers are responsible for arranging their own delivery.

❧ GROWING MEDIUMS ❧

There is a staggering variety of options from which to choose. Most commercial mixes contain peat, perlite, vermiculite, a wetting agent, and a nutrient charge (soluble fertilizers). The vast majority of these cannot be used by certified-organic growers because of the wetting agent and fertilizers. However, there are now several companies that make potting mixes for organic production.

If you are inexperienced at transplant production, start with a commercial mix. Ask the sales rep for the horticulture supply company you use to recommend its most popular mix for growing vegetables. Tell him or her the plug size you'll be using most often, as this can determine how coarse the mix should be. If you are growing only transplants (not hanging baskets, gallon pots, etc.) you should be able to get by with just two kinds of potting mix: a fine seed-starting mix and a somewhat coarser growing mix. You'll also want to buy a small bag of vermiculite; seeds that should not be covered by soil mix will benefit from a dusting of vermiculite, which allows light to reach the seed but retains moisture around it. Over time, you may want to experiment with different brands and grades of mix in response to any problems you may experience.

Many growers, especially organic growers, make their own growing mixes. We did this ourselves for many years, when there were no premade organic mixes on the market. Recipes for farm-made growing mixes are available from ATTRA and the Ohio Ecological Food and Farm Association (OEFFA); see the resource list for contact information. On a small scale, you can mix up the ingredients in a large tub. In larger greenhouses, cement mixers are often used.

Buying Plants

If you don't have a greenhouse, you can direct-seed as much as possible and buy the rest from plant and plug suppliers. Several excellent nurseries stand ready to fulfill your plant orders, but there are several things you need to know before you contact them.

First, most plug suppliers don't sell directly to growers. Instead, they work with brokers. The brokers have sales reps, either field reps or in-house reps, who will take your order and invoice

Storing Leftover Seed

Many types of vegetable seeds remain viable for longer than a year, provided they are stored correctly. Here are some tips:

❧ Don't leave packets open. Seed can absorb 2 percent of its weight in moisture over an hour's time, especially in a humid greenhouse.

❧ Store seed in an airtight plastic container, like the kind you use in the kitchen, and put a ¼-inch layer of silica gel dessicant in the bottom.

❧ The ideal moisture content for seeds is 5 percent to 8 percent. Every 1 percent decrease in moisture level down to the optimum will double the storage life of the seed.

❧ Ideal temperature depends on humidity. The temperature (°F) plus the percentage relative humidity should be less than 100. For example, in a 40°F refrigerator, the humidity should be less than 60 percent.

❧ Seed can be stored as low as 32°F (0°C), and seed of some cool-loving crops can be stored in the freezer.

Under these conditions, here's how long you can expect seed to remain viable:

5 years: Broccoli, Brussels sprouts, cabbage, cauliflower, celeriac, celery, chicory, Chinese cabbage, cucumber, eggplant, endive, kale, kohlrabi, lettuce, muskmelon, New Zealand spinach, radish, rutabaga, spinach, squash, turnip, watermelon

4 years: Beets, peppers, pumpkin, tomato

3 years: Beans, carrots, peas

2 years: Okra, parsley, salsify

1–2 years: Onion, parsnip, sweet corn, Swiss chard.

If you're unsure about seed viability, here's a simple way to test it: Count out a specific number of saved seeds, such as 25 or 50, and put them in a moist paper towel, inside a plastic bag. Put them in a warm place for several days (the usual days required to germinate that species), then take them out and count the number that have germinated. You can then figure out the germination percentage and decide whether it's adequate for your purposes.

In the grand scheme of things, though, seed is cheap. When in doubt, buy new seed.

---

you or process your credit card payment. The broker then places the order with the supplier, who ships directly to you. Several commercial seed companies serve as brokers for the biggest plug suppliers; Gloeckner, Germania, and Harris come to mind first. Gloeckner works through field reps, and Harris and Germania take orders in-house.

The second thing to know is that most plug suppliers require that you have a sales tax exemption form or certificate, which proves you're in business, not just a homeowner who wants to buy plants cheaper than retail.

Third, be aware that plugs are available in many sizes. You might purchase 200s (200 plants per tray), 72s, or even 36s. The smaller plugs are cheaper per plant but more expensive per tray.

Another consideration is whether you can meet the minimum order required by the supplier. Some are picky about how big your order must be, requiring, for example, that you get at least two trays of each variety and that you must order a total of five trays to make a box. That's starting to change, and many suppliers now have only a one-tray minimum. I've heard beginning growers say they don't need a full tray of a variety, but if you can't use a full tray, you're not serious about growing commercially. You should be growing everything in full-tray quantities. Two hundred plants of any crop is not a lot. Even if you don't want to plant them all yourself, you can probably sell the extra plants to your customers. You also can buy larger plants; if you don't need or think you can resell 200 thyme plants, for example, you can probably use 36. Think big!

Finally, when working with plug suppliers, you must plan ahead. Most crops have lead times of eight to 12 weeks, so get your orders in early. If you find yourself short of something, you might get lucky and find a plug supplier with some extras. They call this "availability," so check the website or call and request their availability list for the week you want to plant. Most plant suppliers don't go by dates; instead, they number the weeks in the year, beginning with the first Monday of January, so if you want to plant around May 1, you're probably looking at a week 18 delivery.

Several wholesale plug suppliers are listed at the end of the book. At least one of these suppliers grows certified-organic plugs, and the number is likely to increase. You'll also find contact information for Gloeckner and Germania seed companies.

▶ BARE-ROOT TRANSPLANTS ◀

Some vegetables are widely planted as bare-root transplants. These include onions, sweet potatoes, and asparagus. Strawberries are also sold primarily as bare-root plants, as are many other brambles and tree fruits. When ordering bare-root plants, be sure you'll be ready for their arrival. The roots cannot be allowed to dry out, and they must be planted and watered in as soon after arrival as possible.

In years past, vegetable transplants were grown in the South in sandy soil, then pulled and shipped bare-root to vegetable farms all over the country. Now containerized plugs are more common in the vegetable industry. Some bare-root transplants are still available, but usually only in large quantities, making them unsuitable for the market gardener. The University of California Extension recommends against bare-root vegetable transplants because of the potential for damaging root hairs and spreading diseases when the plants are removed from soil and bundled for shipping.

Hardening Off

Whether you grow your own or purchase transplants from an outside supplier, you must harden off the plants before you plant them outside. Up to this point, the plants have been growing in the most hospitable environment imaginable—the warm, humid, sunny, still greenhouse, where water and nutrients are delivered whenever needed, and the plant is rooted in a friable, perfectly balanced medium. It's a far cry from what awaits in the field—drying wind, harsh sunlight, fluctuating temperatures, less water, less-available fertilizer. No wonder plants often suffer transplant shock.

Hardening off means introducing plants gradually to the harsh realities that await them outside. Gardening books always recommend that you move them outside for a few hours, increasing the length of time every day for a week or more.

But market gardeners don't have time for all that. Most people growing commercially move thousands of plants through the greenhouse each year and can't continually move them in and out every day for a few weeks. So most growers create separate hardening off areas where plants are moved about a week before they are scheduled to go into the field.

The easiest solution is to move trays into an unheated hoophouse with roll-up sides, where the transplants will be subjected to brighter sunlight, more wind, and greater temperature swings, but not quite as harsh as the conditions that would be found in the field. Another solution is to put the trays on pallets or plastic crates outside the greenhouse, with the greenhouse itself providing some wind protection. Or, you can set up a low tunnel with PVC hoops attached to pallets and covered with plastic or row cover.

Some growers use their market canopies to provide some protection during the hardening-off period. Most of the time, the canopies aren't yet needed for market, so there's no rush to move the transplants out from under them.

In climates where freezes threaten at hardening-off time, it makes sense to put the trays someplace where they can be minimally heated. A lean-to structure can be constructed with PVC hoops against the heated greenhouse, and sides rolled up to allow heat into the lean-to.

Even small metal hoops covered with row cover or plastic will suffice to give the transplants a halfway house to ease their way into the field.

Another aspect of hardening off that is often overlooked is fertility. Plants that have been given ample water and nutrients right up to transplant time are going to be soft, and may be set back severely by field conditions. Decrease feeding and watering the week before hardening off; plants that are a bit on the hungry side will respond well to a light feeding two or three days before transplant time and will be ready to grow in the field.

Preparing Beds

On many small-scale market gardens, crops are grown intensively on raised beds. There are many reasons for using raised beds. They allow the soil to drain better, which means you often can get into the field earlier in spring when the weather is rainy. They also allow you to plant more intensively and to target fertilizers directly to the plants.

The easiest way to make raised beds is to have the field plowed and disked, either with your own tractor or by hiring someone else to do it. On a small scale, you can make raised beds with a tiller and a furrower attachment. First, till the soil to a depth of 4 to 5 inches. Figure out how you want your fields to be laid out, based on access and where you will need to run irrigation lines. Use the tiller and furrower and begin tilling a straight row. The wings of the furrowing attachment will push the soil off to the sides; this is where your path will be. At the end of the row, turn around and bring the tiller back parallel and about 4 feet away from your first furrow. Again,

Cover Crops

Cover cropping is essential knowledge for the small-scale, sustainable farmer. Cover crops serve several important functions in the management of the farm: They cover the soil, preventing erosion, during winter and other fallow periods; they increase nutrients, organic matter, and biological activity in the soil; they can help with weed control; they may provide habitat for beneficial insects; and they can be cut and used to mulch other crops. The choice of plants for cover cropping depends largely on climate, season, and the length of time the field isn't needed. Some crops, such as buckwheat, are planted as a quick summer cover, primarily to improve soil structure when turned in, while others, such as hairy vetch, are nitrogen-fixing legumes that are grown over the winter. Most growers put cover crops into their rotations, so that all fields have a period of rest and renewal every few years.

The subject of cover crops is complex because of all the variables, and it is quickly evolving, thanks to ongoing research on soil health. This is one field of study that will make all other aspects of production easier in the long term. You will find an excellent summary of cover cropping from ATTRA at www.attra.org, including a long list of other publications for further investigation.

the furrower will throw the soil up next to the first pile of soil. This will create a raised bed, which will need to be raked or lightly tilled to flatten it for planting.

You also can do this with a two-bottom plow on a tractor. In the same way as described above, plow the length of the field in one direction, which creates a furrow while mounding the soil on the right. Then turn the tractor and plow in the opposite direction, about 6 feet away from the first furrow. The plow pushes the soil to the right again, toward the mound that was made by the first pass. When you have completed the second pass, your first raised bed will be formed. Reverse directions again, driving with the left tractor tire on the bed you just made. You will widen the furrow, making a wider path, and throwing soil to the right, which will become your next bed. To flatten the tops of the beds, you can either disk with the tractor or till lightly. Your beds will be about 4 feet wide on top.

That's the low-cost way to make raised beds when starting out. Later, if you want to invest in time savings, you can purchase a bed shaper for your tractor. This is an implement that raises the soil and flattens it in a single pass. Equipment is also available that lays plastic mulch and drip tape at the same time.

Transplanting

Your choices for planting transplants are pretty simple: You either plant them by hand, or you use a tractor-pulled transplanter.

If you're going to plant by hand—and it's really not that difficult to do an acre or two this way—use a string to mark the rows until you get good enough to plant straight without guidance. Some growers use rolling dibbles that make indentations where the plants should be placed. Straight lines of plants will become tremendously important later, when weeding and laying drip irrigation lines. They also look better and will win you admiring looks from other farmers. Seriously—this is part of the craft.

Our system for hand-planting is to till the soil a couple of weeks before planting, then till again, shallowly, or wheel hoe immediately before planting. This removes the weeds that have germinated since the first tilling, and it creates a crumbly texture that is easy to plant into. Next, we remove the transplants from the plug tray and drop them onto the bed where they will be planted. It's nice to work with a partner, so one person can pull the transplants from the plug trays and drop them onto the bed, while the second person moves along behind, planting them. Trade places after each row or bed, so one person doesn't have to be on his or her knees for long periods. If the soil is crumbly, you can probably just scoop out a hole for the transplant with your hand. If it's hard, you may have to use a hand tool. We find that the best tool for preventing hand strain is a right-angle trowel that lets you pull the soil toward you to open the hole.

When you hand-plant a bed, it's important to come along immediately with the hose to water in the transplants. Even in moist soil, you should water the transplants to set them in the soil and to wash soil up against the roots, eliminating air pockets. Use a long wand with a soft flow breaker and circle the transplant, nearly touching the soil. Take your time and be sure to deliver enough water in a circle around the plant so that its roots will immediately start to move into surrounding soil. The biggest mistake we see inexperienced planters make is holding the hose wand several feet above the plants—a bad practice because it doesn't deliver enough water to the soil immediately around the transplant and it waters a broader area where weed seeds are just waiting to germinate.

If you're planting more than an acre or two of transplants, consider investing in a mechanical transplanter. These implements are pulled behind the tractor; they have seats where workers sit and remove transplants from the trays and drop them down a planting tube. The implement opens a hole, drops in the transplant, closes the soil around it, and (with some models) gives it a drink of water. A transplanter requires a tractor with a creeper gear and at least two people—one to drive the tractor and one to sit on the transplanter dropping the plants. Depending on the brand of transplanter, four or more units can be connected to plant multiple rows in a single pass. Some transplanters will plant through plastic mulch. To get a sense of the wide range of transplanters available, plan to spend some time on the Internet looking at the websites for new and used vegetable equipment dealers. It's a lot to digest at first, but if you allow yourself plenty of time to peruse the sites carefully, looking at all the photos and reading the descriptions, you'll soon feel a lot more knowledgeable about equipment. Just do a search for "mechanical transplanter" and go where it leads.

❧ EARLY-SEASON CROPS ❧

Some growers get a jump on the season by planting under low tunnels, which act like mini greenhouses in the field. First, you need drip irrigation on the bed, then a layer of black mulch film (see section on mulches, below) to prevent the growth of weeds. Then the transplants are planted through the plastic mulch. Then wire hoops are set up every 4 feet along the bed. These hoops are then covered with perforated poly row cover or frost fabric. The edges of the row cover must be secured and tied tautly at the ends.

Other season-extension strategies include growing in unheated hoophouses, growing on black plastic or infrared-transmitting (IRT) mulch, and growing under floating row cover. These are all covered in separate sections.

❧ DIRECT-SEEDING ❧

Some crops, such as carrots and beets, must be direct-seeded. Some, including lettuce and sweet corn, are usually direct-seeded but can be started in the greenhouse and transplanted out. Others (tomatoes, for example) are rarely direct-seeded. You'll have to learn what works best in your soil when you have a choice about direct-seeding or transplanting. In general, having fine-textured soil and keeping it moist will help seeds germinate in the field.

You can sow seeds by hand in a furrow, or you can use a push seeder, such as the Earthway, when growing on a small scale. For larger farms, precision seeders are more appropriate. In either case, plant in straight lines to allow for easier cultivation later. See chapter 4 for a more complete discussion of seeders.

Weed Control

Every crop has a critical period in which it needs to be weed-free for best production. This is generally about one-fourth to one-third of its life in the field. After that, plants have become established and are better able to compete with weeds. That doesn't mean you shouldn't keep your crops weed-free as long as you possibly can; it's just not as critical toward harvest time. Big weeds can cause other problems besides competition for water and nutrients—they can harbor insect pests, and they can make the farmer's work unpleasant. They don't look good, either.

Farmers employ several strategies for keeping fields weed-free, depending on whether the plants are direct-seeded or transplanted.

➤ STALE SEEDBED ◄

The key to success in direct-seeding is to have a weed-free seedbed so your crop seeds can germinate and get established without competition from weeds. If you know there are a lot of weed seeds in your soil, you might want to employ a strategy known as the stale seedbed. The idea is to till the soil, then irrigate to get the weed seeds in the top few inches to germinate. When the weed seedlings are just 1 to 2 inches tall, you must eradicate them.

Some growers weed the small seedlings with a skimming hoe that barely goes beneath the surface. You don't want to hoe the soil too vigorously, and you definitely don't want to till it again because you'll just churn up more weed seeds, which will then germinate later.

Flame weeding is the most effective way to achieve a stale seedbed, and it's highly recommended for crops that can't compete with weeds, such as carrots and onions. A flame weeder is a piece of equipment that directs propane-fueled flames onto a crop bed. Many models are available, both handheld and tractor-mounted. Prices range from a few hundred dollars for one or two burners to several thousand for multiburner flamers mounted on toolbars. Flame weeding doesn't actually burn up the weed seedlings; it just heats them to the point where their cells burst. The weeds won't look dead until later.

Even in a stale seedbed, direct seeding can disturb the soil enough to bring weed seeds to the surface, so some growers practice what is known as peak emergence technique. This involves tilling and planting at the same time, then flame weeding a day or two before the crop seeds germinate. How do you know when that is? You can either dig up a few feet of the seeded row to see if the seeds are sending up shoots. Or you can cover a section of row with glass or plastic; the crop seeds will come up under the plastic a day or two before the uncovered seeds. As soon as the seeds germinate beneath the plastic, it's time to flame-weed. By the time the weeds shrivel and die a day or two after flaming, the crops will be coming up into a weed-free row.

➤ CULTIVATION ◄

Weed seeds will continue to germinate all season long, some in response to changing temperatures and moisture, others in response to physical disturbances that expose them to light. Most sustainable farmers deal with weeds with a program of regular cultivation.

The best time to deal with weeds is when they are just an inch or two tall. There are many kinds of tools designed for weed control in vegetables. For the small-scale market gardener, it's the wheel hoe. On a larger scale, there are countless implements to pull behind the tractor. The best detailed discussion about all this equipment is a book called *Steel in the Field: A Farmer's Guide to Weed Management Tools* from the Sustainable Agriculture Network. There's also a great video called *Vegetable Farmers and Their Weed Control Machines*. See the resource list for ordering information.

➤ MULCHES ◄

Three categories of mulch are used by vegetable farmers: plastic, paper, and such organic materials as straw and hay.

Mulches prevent light from reaching bare soil, thereby preventing the germination and growth of weeds. They have the added

Plastic mulch has to be laid in tight contact with the soil; it's best accomplished with a bed shaper–mulch layer implement for a tractor.

benefit of preserving soil moisture, and organic mulches also contribute organic matter and nutrients to the soil as they break down over the course of the season. Each type of mulch has its disadvantages, though.

Polyethylene mulches are favored by large-scale vegetable farmers because they can be laid down quickly and efficiently with tractor-pulled implements. They are the least expensive mulches to purchase.

Plastic mulch is available in several colors that have different effects on soil and plants:

- **Black mulch** warms the soil as much as 5°F (about 3°C) at a 2-inch depth on a sunny day, provided it is in tight contact with the soil. It is generally used for early plantings and heat-loving crops.
- **White-on-black mulch** causes a decrease in soil temperature, because the white surface reflects the sun's energy back into the plants' canopies. The black backing prevents light from reaching weed seeds.

- **Silver reflective mulch** is used to repel aphids, which seem to be confused by it.
- **Infrared-transmitting (IRT) mulch** can be brown or green; it warms the soil more than black mulch.
- **Red mulch** has been shown to benefit tomatoes by increasing yields and reducing early blight. Other studies show increased yields in zucchini and melons.
- **Other colors** under investigation include yellow, blue, gray, and orange, but so far there are no firm recommendations on how and when they should be used.

The downside of plastic mulch is that it cannot be recycled in most parts of the country and therefore is a huge waste problem. There is also a significant amount of labor that goes into removing the mulch from the field at the end of the season. Organic growers are allowed to use plastic mulches, but they must remove them each year.

Another type of mulch that is useful in some situations is landscape fabric, a heavy, black, woven polypropylene material. There are several brands and grades; the heaviest will last 20 years or more. Although it is most often used on perennial beds to control weeds, some growers use it for annuals and just roll it up and store it at the end of the season. It needs to be cut with a sharp kraft knife or melted with a propane torch or even a branding iron to create openings for plants.

Paper or other biodegradable mulches have been created in response to the disposal issues surrounding plastic mulch. In theory, these mulches will remain intact for a month or two and then begin to degrade, so that they are virtually gone by the end of the season. Several products are available, and this seems to be an area of interest for manufacturers, so check the Internet for the latest innovations. At this writing, at least two products are approved for organic farms. If you are certified, either look for the Organic Materials Review Institute (OMRI) label or check with your certification agency.

Natural mulches, such as straw, hay, wood chips, and grass clippings, benefit the soil (provided they are not contaminated with herbicides; see the sidebar on clopyralid). They conserve moisture, and they help keep soil cool in summer. The downside is that they have to be applied by hand. At Wheatland Vegetable Farms in Purcellville, Virginia, Chip and Susan Planck mulch every crop on their 25 producing acres. They have a neighboring farmer make round bales that are the right size for their farm—4 feet wide and weighing

500 to 700 pounds. At that size, the bale can be unrolled by one person and it will cover everything but the crop in their 6-foot-center beds. At other farms, rectangular bales are moved to the row ends by tractor and then broken apart and spread by hand.

Some of the best hay for mulching vegetables is alfalfa hay because it is high in nitrogen, which feeds the crop as it breaks down. Alfalfa is usually free of weed seeds, unlike some fescue or prairie hay.

One precaution about using any kind of baled hay or straw: wear a breathing mask when moving and spreading it. Breathing moldy hay or straw is the cause of farmer's lung, a respiratory disease that can cause permanent lung damage. Hay and straw often heat up after they are baled, and mold develops inside the bale. As the spoiled hay dries, it darkens and crumbles easily to an extremely fine dust. People can breathe these dust particles into the innermost regions of the lungs, the alveoli, which can get inflamed.

Biodegradable film mulch is used on the beds, and the paths are mulched with clean straw on this well-kept field.

Facts about Persistent Herbicides

What is it? Clopyralid (clo-PEER-a-lid) is an herbicide that is widely used in agriculture and lawn care. Other herbicides that can be equally problematic include picloram and aminopyralid.

Brand names? In lawn care: Confront, Lontrel, Momentum, "Weed & Feed" fertilizers, possibly others. In agriculture: Curtail, Hornet, Milestone, Redeem, Stinger, Transline, WideMatch, possibly others.

What's the problem? Clopyralid does not degrade when composted, as most other agricultural chemicals do.

How long can it last? The half-life of clopyralid in compost is greater than one year.

What are its effects? Clopyralid affects numerous horticultural crops. Plants damaged by clopyralid will show:

* Stunted growth (the main growth tip stops growing and the lateral buds begin to grow)
* Reduced fruit set
* Cupping of leaves
* Failure of secondary leaves to grow after the seed leaves emerge
* In legumes, compound leaves stay single.

What can you do? Don't use products containing clopyralid and other persistent herbicides on your property. Don't buy hay or straw that may have been treated with it. Don't use grass clippings from lawns where these products may have been used.

For more information, see the North Carolina State University fact sheet *Herbicide Carryover in Hay, Manure, Compost, and Grass Clippings*. www.ces.ncsu.edu/fletcher/programs/ncorganic/special-pubs/herbicide_carryover.pdf.

Insect and Disease Control

One of the most challenging aspects of diversified market farming is keeping your crops healthy. There are two keys to this: First, when you grow a plant under optimum conditions, you are less likely to have problems with insects and diseases. Optimum conditions include choosing appropriate varieties, planting at the right time, planting into healthy soil, using the best spacing, keeping the young plants weed-free, and providing water and fertility as needed by the plant. There is ample scientific evidence that well-grown crops are less attractive to insect pests and more resistant to diseases.

The other key to insect and disease control is to know and understand the life cycles of each. This is a tall order, when you consider that you might grow 20, 40, or more different crops each year, and each crop has several insects and diseases that can harm it. It becomes even more complicated when you realize that beneficial insects, naturally occurring on your farm, are often the best solution to pest problems. As an ecological farmer, you must recognize the good guys, the bad guys, and when you can let the former take care of the latter and when you must get involved. Once you have determined that you have to take action to save a crop, you must know which methods and products are safest and most appropriate.

Many gardeners assume there is some easy answer to every pest problem, and for those who are not interested in ecological farming, there probably is a broad-spectrum pesticide that will kill everything moving in the garden. In addition to their negative effects on both human and environmental health, broad-spectrum pesticides are not a long-term solution to pest problems. Pests will bounce back as soon as the residual activity of the pesticide declines, and they'll be more abundant than before because beneficial insects also will have been killed. Using broad-spectrum pesticides puts you on a treadmill that is hard to get off.

The most successful farmers are those who can identify insects on their crops, understand where in the life cycle that insect is, and know what ecological controls are available to prevent the insect from ruining the crop. This is a field of study that takes many years, but it is a critical part of the farmer's homework. My advice is to get a good insect identification book and a guide to ecological or least-toxic controls, and then monitor the insects on your farm carefully. My current favorites are *Garden Insects of North America* for identification and *Pests of the Small Farm and Garden* for identification and controls. For certified organic farmers, a third essential is the list of approved pest-control products provided by your certification agency or by the Organic Materials Review Institute (www.omri.org).

Products, however, are not the only way to approach pest control. In some cases, the first line of defense may be as simple as scheduling crops to avoid the presence of the most damaging pests. Many insects develop in response to environmental factors, including day length and temperature. By careful observation and recordkeeping, the farmer can determine when certain insects become a problem and how long they are present. In some cases, crops can be planted earlier or later to avoid a specific insect pest.

A second approach is to use physical barriers to protect crops during vulnerable times. The most common barrier is row cover

Managing Row Cover

Row covers are a great benefit to market farmers, offering protection from insects and frosts, and providing a warmer environment for early crops. Growers have to deal with three main problems in using row covers effectively.

First, row covers need to be anchored so that wind doesn't remove them from the crops. There's nothing worse than having a strong wind lift up a hundred dollars' worth of row cover and wrap it around a tree high

Three ways to anchor row cover: with plastic staples sold specifically for the purpose (as shown at the Rodale Research Farm in Pennsylvania), by burying the edges (at Johnny's Selected Seeds in Maine), or with sandbags (at Meadowlark Farm in Michigan).

made of lightweight, spun-polypropylene fabric. Row covers are made by several manufacturers and available from most farm supply companies. They are available in several weights, for use in different seasons. In general, there is a tradeoff between frost protection and light transmission, so you'll want to choose row cover carefully to suit your needs.

overhead. Not only is the row cover lost, but you'll also have to look at it for years. To anchor row covers, growers use whatever is on hand and heavy enough to withstand strong winds. Sandbags are a great solution, and you can also purchase net bags to fill with rocks if you happen to have rocky soil. Sections of hose filled with water and clamped shut may be heavy enough in some areas. Some growers bury the edge of the row cover, but that shortens its life. Lengths of rebar or t-post can be used. Plastic stakes designed for the purpose are available commercially.

The second management issue is how to lift the row cover above the crop so that the growing tips of plants aren't damaged. Although row cover can be "floated," or laid directly on top, of some plants, many types of plants can be damaged by having row covers rubbing on them. Peppers and tomatoes are good examples of plants that have growing tips at the top of the plant, and they will do better if the row cover is suspended above the growing tips.

The best way to accomplish this is with wire hoops or wickets. Hoops are easy to make from #9 galvanized wire. You'll have to experiment with the length of the wire, as it depends on the width of the bed and the height of the crops. An 8-foot length will give you a hoop with 12-inch legs that you can push into the soil to anchor it firmly. Hoops can be set on the bed every 4 to 6 feet, then covered with row cover. For extra stability, place a hoop above the row cover at every other hoop.

Another great resource is the wire wickets from campaign signs after an election. Many signs never get picked up or are discarded by the candidates. The wire frames for the signs can be reused to form flat-topped tunnels for low crops like lettuce and spinach.

Finally, there is the issue of storing row cover. It should be protected when not in use to prolong its life. Once it's been unrolled, it can be unwieldy. Some growers use lengths of PVC pipe to roll the row cover on, then label the pipe with the length of the piece and any other pertinent information. Others just fold it up and stuff it into grain sacks or plastic bags and hang it from the rafters in the barn to prevent mice from nesting in it.

Don't throw out row cover just because it gets a few holes. It can still come in handy. Deer won't eat lettuce through row cover, even if it's torn up. And it's convenient to have smaller pieces in the greenhouse to place over flats when germinating seeds.

Some of the brand names available in the United States are Reemay, Typar, Agribon, Agro-Fabric, and Covertan, which are all lightweight, spun-bonded fabrics. There is also a Japanese product called Tufbell, made of polyvinyl alcohol.

The lightest row covers weigh less than half an ounce per square yard and are used primarily to exclude insects in summer. Moving up from there, weights can range from ½ ounce to 2 ounces per square yard. As the weight increases, the level of frost protection increases, but light transmission decreases.

Tufbell is the exception, giving 10°F (about 5.5°C) frost protection, while still allowing 95 percent light transmittance. It is the most expensive by far, costing about 20 cents per square foot compared to the 1.5 cents for lightweight row covers. Growers who have used it, though, say that it can last five years or more when cared for properly.

❧ DEGREE-DAYS AND ❦ INSECT DEVELOPMENT

Observant gardeners have always known that some events occur simultaneously every year, no matter what the weather or date. The lilacs are always blooming when the maggots first appear on cabbages; you'll find squash vine borer eggs when the roadside chicory is in bloom. These correlations aren't just folklore. They are actually the result of parallel development of certain insects and plants based on the amount of time both have been able to grow and develop. The concept is known as growing degree-days (GDD), and it is an important part of integrated pest management (IPM), a least-toxic system of managing pests by monitoring for them and taking action only when they reach certain thresholds.

Farmers can use growing degree-days to predict and avoid insect problems and to plan harvests. GDD takes into account the fact that insects, because they are cold-blooded, depend on environmental temperature for development. The same is true of many plants, although some depend on day length more than temperature. For each day that the average temperature is one degree above the base temperature at which growth begins, one degree-day has accumulated. For most plants, the temperature at which growth starts is between 45°F and 55°F, so the base point of GDD calculations has been set at 50°F.

The formula for determining degree-days is this:

$$\frac{\text{Maximum temperature} + \text{minimum temperature}}{2} - \text{base temperature} = \text{GDD}$$

For example, if on March 3 the maximum temperature is 60°F and minimum temperature is 50°F, here's how to figure degree-days:

$$60 + 50 = 110$$
$$110 \div 2 = 55$$
$$55 - 50 = 5 \text{ GDD}$$

If the average temperature is equal to or less than the base temperature, no degree-days are accumulated. If the temperature is higher than 86°F, the maximum is registered as 86°F.

Growers interested in using GDD should get a maximum-minimum thermometer and check it every day, rather than relying on published National Weather Service reports, because temperatures on a farm will vary substantially from those at official weather stations. Start recording GDD as soon as the temperature reaches 50°F (10°C)—between January 1 in the South and March 1 in the North.

Once growers have started recording GDD, they should look up GDD data on insects that are often a problem for them. Only insects that overwinter in their area should be considered, however, because migrating insects are dependent on wind patterns rather than heat accumulation. Also, insects that have only one life stage present at a time should be considered; aphids, for example, can be present in all life stages simultaneously, so GDD can't be used to predict their appearance.

In general, the major vegetable pests that can be predicted using GDD include Colorado potato beetle, corn rootworm beetle, European corn borer, cabbage maggot, onion maggot, seed corn maggot, and squash vine borer.

Information on GDD is available at the University of California's IPM website: www.ipm.ucdavis.edu. The site has plenty of information about the general practice of using GDD to predict events, plus it has a phenology database with GDD on many insects. The GDD for a specific pest will be valid whether those insects are in California, Maine, or anyplace else, as they always develop at the same rate. The GDD equation for Celsius and other information can be found at the Ontario Ministry of Agriculture website: www.omafra.gov.on.ca/english/crops/pub811/10using.htm.

Potato growers, for example, can be ready to spray with the bacteria Bt when 185 degree-days have accumulated. That will be the time of the first instar of the Colorado potato beetle. Or growers might want to change planting dates to avoid a particular pest, such as the cabbage maggot, planting either well ahead and protecting with row cover or waiting until the egg-laying phase has passed.

GDD information is plentiful for nursery crops because an Illinois nursery inspector spent 25 years recording and correlating the development of trees and shrubs and their insect pests. Vegetable growers are beginning to do the same, and the science of GDD is increasingly a useful tool for ecological farmers.

Irrigation

If you're a gambler and you don't mind throwing money away, you can try to grow vegetables and flowers without irrigation. But if you really want to earn some money for all your work and expense, be sure you have a way to irrigate. No matter where you live, no matter how much rain you usually get during the growing season, you need to be able to irrigate your crops. Vegetables are 80 to 95 percent water, so their yield and quality is very quickly affected by a lack of water. In some quick-draining, sandy soils, plants may need to be irrigated three days after an inch of rain, especially if they are in a critical development period. Shallow-rooted crops should never be without water for seven to 10 continuous days. In general, most vegetables and flowers need 1 to 1½ inches of water per week.

The most common irrigation methods for vegetables are sprinkler, or overhead, irrigation and drip, or trickle, irrigation. Some farmers also use flood irrigation, but it's a small percentage, and it's used only in a few areas of the country where ditches and gates are already commonplace.

☞ OVERHEAD IRRIGATION ☜

Water can be delivered to vegetable crops on a small-scale vegetable farm by small, low-volume overhead sprinklers that are either hand-moved or solid-set. Penn State University Extension estimates that a hand-moved sprinkler system costs $2,000 to $3,000 per acre. A solid-set system, which is permanently installed in the field, costs about $3,500 per acre. What's known as "big guns" are high-volume sprinklers that can cover more area. Their cost can range from $5,000 to $15,000 per acre. These systems are obviously better suited to market farms on the 3+ acre scale. For smaller market gardens, water reel traveling sprinklers are often used for overhead irrigation. There are also mini sprinklers, which are used to keep specific areas moist, such as beds of lettuce.

The disadvantages of any kind of overhead irrigation, though, are numerous. A large percentage of the water applied is lost through evaporation or blown off course by the wind. The amount of water delivered is irregular throughout the field. With high-pressure irrigation, droplet size can be big enough to damage the growing tips of young crops. Wet foliage can lead to foliar diseases. Soil splash can spread soilborne diseases.

➤ DRIP IRRIGATION ◄

The most efficient and popular way to deliver water to crops is by drip irrigation. Compared to sprinkler irrigation, drip uses 20 to 40 percent less water and can be used on windy days and on any type of topography. Besides saving water by delivering it directly to the plants' root zone, drip also helps prevent foliar disease in crops and reduces the weed growth outside the crop row. It can be used to deliver fertilizers, known as "fertigation." Drip irrigation is possible with very low water pressure, so it may be the only way to irrigate in certain circumstances. The downsides of drip are that setup requires more intelligence and attention by the farmer and that it

Drip irrigation delivers water where it's needed and no place else, saving on water and weeding. The flexible tape can be used on contoured fields as it is here, at Meadowlark Farm in Michigan.

Drip tape is attached to a header hose with a twist-lock connector. All components of drip irrigation systems are readily available from numerous suppliers.

creates a waste product (the tape) that can't be recycled in most areas of the United States.

If you go online and do a search for "drip irrigation vegetables," you will find numerous university publications. Most of these can be intimidating, with diagrams of complicated-looking systems and lots of talk about valves and pounds per square inch (psi) and water-holding capacity. You may feel unqualified to set up a drip system after reading some of this, but rest assured that there is more information there than a beginner needs to know. I'm going to tell you how to set up the simplest possible drip irrigation system. Once you see how easy it is, you'll be thrilled. By careful observation of your simple drip system, you will learn whether you need to get more sophisticated, but chances are good that this easy method will do the job.

All you need are drip tape with emitters every 8 to 12 inches; one twist-lock connector for each run of drip tape you plan to use; a hole punch; a ½-inch or ¾-inch black poly hose to run across the top of five or six beds; compression fittings to start and end this header hose; a filter and pressure regulator; and a hose to get water from your water supply to the field.

Drip tape is available in many thicknesses, described as "mils." (A mil is ¹⁄₁,₀₀₀th of an inch.) You can get 6-mil tape or 12-mil tape or various other strengths; the larger the number, the more expensive it is. We haven't found any advantage to buying the heavier stuff. If you are careful with the tape, it lasts equally long at any thickness. If you are not careful, you're as likely to cut up thick tape as thin tape. So start with the least expensive, and see if it fits your needs.

◈ LAYING OUT A SYSTEM ◈

Here's how to lay out a simple and inexpensive drip system:

- First, lay out the ½-inch black poly main line that will deliver water to the drip tape. We run it across the top of the beds but within the cultivated area so we don't hit it with the mower when mowing the grass path at the top of the field. The number of lines you can irrigate at one time depends on the length of your beds, your water pressure, and emitter spacing. Ask the irrigation supplier when you buy your drip tape for a recommendation. Our beds are 125 feet long. We run two lengths of drip tape per bed, in a block of five beds. So each block has about 1,250 feet of drip in place.
- Use a compression fitting with a female hose beginning at one end of the main line and a male hose end with a cap at the other. The screw-off cap allows you to flush the line.
- Lay the drip tape in straight lines on your beds. To unroll the tape from the big reel it comes on, we put a rod through the center of the reel and suspend it over the garden cart. Then we grab the end of the tape and walk all the way down the bed. If it's windy, we will tack the drip tape down with an earth staple (a U-shaped piece of wire), though not too tightly. Use two lines of drip tape in a 4-foot-wide bed. Cut the tape straight across, not at an angle, when you have unrolled enough to cover the length of the bed. At the end closest to your water supply, you will connect the drip tape to a header line; at the other end, you should tie a knot in the drip tape to close off the line.
- Punch holes in the main line where the drip tape will connect. You will need twist-lock connectors with barbs and a hole punch in the same size as the barb. Punch the hole in the main line and insert the barb end of the twist-lock connector. Insert the other end into the drip tape and screw the outer ring tight onto the tape.
- Attach a pressure regulator to the female hose end at the beginning of the main line. Then attach the hose from your faucet to the front end of the pressure regulator. Turn on the water, and

Deer Control

Throughout North America, deer have become one of the most troublesome pests on small farms. Deer populations are growing rapidly, and habitat loss due to development is forcing them onto remaining farmland.

Ask 100 farmers what they do about deer and you're likely to get 100 different answers. In general, deer management strategies fall into five categories, each one more difficult and expensive than the one before. In each of these categories, there is one common theme: Do it before the deer find out that you are growing vegetables. Once they know how tasty your crops are, you will have a much more difficult time keeping them away.

❁ **Repellents.** You can make your own repellents from garlic, soap, pet or human hair, fabric softener sheets, or anything else that smells bad to a deer. You also can purchase commercial repellents, which should smell and taste bad to the deer. Read the label carefully, though, because some are not approved for use on food crops. Organic growers may not be able to use even those that are labeled for food crops.

❁ **Scare devices.** Dogs can do a good job if they are left outside at night. Wildlife management agencies sometimes rent gas exploders that detonate at irregular intervals, or you can purchase one for less than $500. The element of surprise is crucial, because once deer have figured out a pattern, they are no longer frightened.

❁ **Barriers.** In cases of sporadic or light deer presence, you may be able to deter them simply by putting something between the crop and the deer. Growers report success stringing yellow caution tape around the garden. Row cover often works. A circle of

you're in business. Once in place, you can easily irrigate a zone by connecting the water to it and going about your other chores.

❁ Check back in about 15 minutes to be sure nothing has gone wrong. Sometimes the drip tape pulls off the twist-lock connectors once the system is pressurized, for example, or the hose might be twisted and not delivering the water to the system. When you first start using drip, check back every hour or so to determine how long you need to run the irrigation. Once you know, you can use a timer on the system to turn it off. Otherwise, think of some way to remind yourself that the irrigation is running, and use it every time, or you run the risk of leaving it running way too long, wasting water, and flooding your field. Our system is goofy but foolproof: We have a string of cheap plastic Mardi Gras beads that we wear when we turn on the irrigation. The necklace is just obnoxious enough that it's impossible to ignore.

posts a foot apart can prevent them from rubbing their antlers on young trees.

- **Baited electric fence.** One or two strands of electrified fence wire or polytape should be marked with flags or strips of aluminum foil spread with peanut butter. Curious deer come up to the fence to investigate and get a small shock. This approach is basically a way of training deer to stay away. You have to check it all the time to be sure it's operating, because if deer get past it one time, they will know to simply jump the wire next time.
- **Fencing.** Many growers report success with 8-foot black plastic mesh deer fencing. The mesh fencing can be stapled to trees or posts, and it blends into the landscape. A 3-D fence consists of a single electrified wire around the outer perimeter of the field and, about 2 feet inside it, a

6-foot-tall, three-strand nonelectric fence. Deer usually won't try to jump two fences. A seven-wire electric fence that slants outward often disorients them and keeps them away. Finally, if all else fails, some growers resort to a 10-foot woven wire fence.

Wildlife management experts say that the only 100 percent effective solution is to reduce deer populations by hunting, and especially by killing does to reduce the herd size over the long term. If that's not your style, be aware that deer are likely to flee to your safe haven during hunting season, so be ready to protect your trees and other crops. Some growers even plant a trap crop that will keep the deer well fed so they won't come looking for more.

Check the resource section at the end of the book for publications on deer control and suppliers of deer fence and repellents.

The costs of drip irrigation will vary considerably based on several factors: the number of lines of drip tape you use per acre, the distance from your water source to your fields, and whether you need a pump to deliver the water or a sand filter to clean it. As a general rule of thumb, though, you can count on spending about $250 per acre for just the drip tape and the connectors when growing on 4-foot beds, with two rows per bed.

Drip tape management

As much as we love drip irrigation, we have to admit that working with it can be a pain. It has to be removed for mechanical weeding, even with a hand hoe, because it is so easily nicked. It gets holes in it, especially from thirsty rodents, that will need to be fixed. And it has to be taken out of the field in winter and stored elsewhere if you hope to get more than one season out of it.

Moving drip tape is a simple matter of pulling it out of the bed when getting ready to weed. In theory, at least, you won't need to weed when the crop gets too big to easily remove the drip tape. Just don't forget it when you go into the field with tractor, tiller, mower, or hoe.

The only possible solution to thirsty rodents is to put pans of water under an emitter every 25 feet or so. In theory, rodents will go for the easy water before they go to the trouble of chewing through drip tape. It doesn't always work in practice, though.

If you do get holes, you have to repair them, because a hole causes the system to lose pressure and results in uneven watering. Irrigation suppliers sell connectors, similar to the twist-lock connector used at the main line, but with the twist mechanism on both ends. To use them, just cut the drip tape on both sides of the hole, then attach the connector and screw the rings tight on both pieces. That works great on small acreages, but if you have a lot of drip tape in the field, you might want to try splicing the tape with the help of a 3-inch piece of ½-inch CPVC tube. Once the tube is inside the tape, attach both sides with a short piece of concrete tie-rod wire, twisted tight with pliers.

You may be able to save your drip tape from one season to the next. Some growers rig up spools and roll the drip tape back onto them. Others just gather it up and put it into a plastic bag that is labeled with length and location. The bags can be left at the end of the row or moved to a storage place.

CHAPTER 7

From Field to Market

Having great produce is essential to your success as a market gardener, but growing is only half the job. As a market farmer, you still have all the selling ahead of you, but even that isn't the end of your work. You still have five additional skills to master: food safety, postharvest handling, value-added processing, pricing, and presentation.

Food Safety

After several widespread produce contamination outbreaks that killed dozens of people, Congress passed the Food Safety Modernization Act (FSMA), which was signed into law in January 2011. In January 2013, the Food and Drug Administration (FDA) published proposed regulations that would require stringent food safety procedures and inspections for farmers. However, the law provides an exemption for small farms that meet these criteria:

* Farms with less than $25,000 a year in food sales are exempt.
* Farms with less than $500,000 annually in gross sales are exempt if more than half of their product sales are to qualified end users, defined as consumers or restaurants and retailers either in-state or within 275 miles of the farm or facility.

At this writing, the proposal is still in the comment stage, and there will probably be several years of phasing in the rules once they become final. But it's safe to say that as things now stand, most

An Online Food Safety Assessment

Complying with new food safety standards is no easy task—there really is an enormous amount of planning, training, and record-keeping involved. But a new, free online tool is making it easier and less overwhelming to work your way through the process of getting your small farm food safety certified.

The On-Farm Food Safety Project (www .onfarmfoodsafety.org) is a web platform that allows anyone to create an account and then answer questions one by one about the multiple aspects of food safety relevant to fruit and vegetable growers. As you answer the questions, the system prompts you with examples of correct responses, based on best practices. As you work your way through the questions, the program collects your answers into a food safety manual. In the front of the manual are your acceptable responses, laid out in a standard food safety manual format. At the end are all the response that were not

acceptable, which gives you a list of issues you need to work on. Once you have created an account, you can save your work at any point and return to it later to continue or correct previous answers.

Make no mistake: It will still take many, many hours to complete the manual, especially if you are new to food safety standards and need to learn all the best practices. But this program is really well designed to provide all the information you need to comply with Good Agricultural Practices. And the final product is a complete manual that you can submit to a food safety auditor when and if you decide to get your farm inspected and certified.

The system lets you know when your answer will cause you to fail an audit. For example, in the section on worker health and hygiene, you will be asked, "Have all your employees been trained in the proper use of toilet facilities?" If you click the button for

people reading this book are going to be exempt from the FSMA rules. That doesn't mean exempt from food safety, however. Everyone in the business of growing and selling food should commit themselves to producing food that is safe, wholesome, and nourishing. To grow food that makes customers sick or worse would be devastating emotionally, as well as financially. I urge you, whatever your scale or level of expertise, to learn about food safety and to correct any deficiencies that exist in your practices. You should begin by reading the standards known as Good Agricultural Practices (GAPs) and Good Handling Practices (GHPs). They are largely commonsense recommendations for preventing contamination of food in the field, packing shed, and markets.

There are many potential sources of contamination on a produce farm: soil; irrigation water; manure or improperly composted no, a popup window opens that says: "Are you sure your answer is no? All employees and visitors must be trained in the proper use of toilet facilities. A NO answer to this question is out of compliance with best practices in food safety. This non-compliance will be reflected in your food safety plan, which may negatively impact your ability to gain entry to wholesale and retail markets and/or gain certification in Good Agricultural Practices."

Other issues that aren't as critical result in more gentle guidance. For example, the next question is: "Do you have a written policy covering protective clothing requirements (including hair covering, jewelry, and artificial nail restrictions, if any)?" If you click no, further information appears about what your policy ought to include, such as requiring hair coverings in the packing shed and prohibiting artificial nails in the field and packing shed.

Another great feature is that, for standards that require record-keeping logs or other forms, there is a link to a template that you can download or print. You don't need to create your own forms. One of the most complicated aspects of GAPs certification is traceability, having a system in place that allows a buyer to trace back to the farm in case of a food safety issue or allows the grower to locate any sold product that is found to be problematic. There are many ways to accomplish traceability, and the program provides multiple examples of systems using stickers, codes, or stamps.

The program is sophisticated and intuitive. For those who need to get food safety certification to sell to schools, supermarkets, restaurants, and other buyers, it is a tremendous resource. And even those who don't want to get certified would do well to peruse the questions and recommendations for potential problems.

compost; wild and domestic animals; field worker hygiene; harvesting equipment; transport containers; wash water; unsanitary handling during packing; equipment used to soak, pack, or cut produce; ice; hydrocoolers; transport vehicles; improper storage temperatures; and cross-contamination in storage, display, or preparation. Review these potential weak links in the chain of your production, harvest, washing, and transportation of produce to determine whether your practices need modifications.

Despite the exemption, many small growers may find that buyers want them to provide assurance about their practices. At the least, buyers may ask to see a food safety plan; at the most, they may require GAPs certification by a third-party agency. Even if no one is requiring you to do so, you might as well learn the recommended practices and create a food safety plan for your farm. It can't hurt, and it will put you much farther down the road to compliance should you ever need it.

Food safety resources for small farmers are abundant and increasing every month. Workshops, webinars, self-assessments, and publications are readily available on the web and from your state Extension service. A great resource for growers is available from Cornell University. Called *Food Safety Begins on the Farm*, the 30-page guide explains the points in the production chain where fruits and vegetables can get contaminated and offers suggestions for preventing contamination. It is available on the web at www.gaps.cornell.edu.

Although some farmers resent any oversight of their practices, the vast majority are willing to learn and comply. And, frankly, food safety is one area where cooperation is essential for the greater good. If one farmer at a farmers market brings food that sickens or kills customers, every vendor at that market will be affected. Furthermore, large growers are already pushing for Congress to remove the exemption for small farms, arguing that contamination is just as likely on a small farm as a large one; a few incidents of contamination from small farms could result in greater regulation for all.

Harvest and Postharvest

You can grow the most beautiful produce on earth, but if you can't get it to market in perfect condition, all your efforts are in vain. The practices collectively known as *postharvest handling* can make the difference between success and failure in market farming, so it's important to pay careful attention to what happens after your produce leaves the field. You need to focus on these matters: harvest containers, cleaning, cooling, packaging, and transportation. Let's consider them one by one.

❧ CONTAINERS ❧

To comply with food safety regulations, containers must be sanitized before you use them for harvest or storage, so most growers use food-safe crates and boxes specifically made for produce or other hard plastic containers that can be washed.

At Hearty Roots Farm in Tivoli, New York, Buckhorn flip-top totes have been power-washed and stacked to dry before harvest. Photograph courtesy of Benjamin Shute

Match the container size to the mode of transport from the field; that is, if you have to hand-carry your harvest, use small, lightweight containers; if you can drive a pickup truck into the field, use stackable boxes to maximize space. Scale is the issue here, so think through your system before you invest in harvest containers.

Consider how you are going to move harvested produce from field to barn. Many growers set up their fields so they can drive a pickup truck along an edge to minimize the hauling distance for produce. This is really important if you're growing heavy produce, such as watermelons and winter squash. Tractor-pulled wagons can serve the same purpose without requiring a road. On a very small market garden, you can use a garden cart. In any case, I advise you to think about this carefully. Don't strain your back any more than is absolutely necessary, or you'll soon be out of business.

Bulb crates or bread crates are among the cheapest harvest containers for market growers. These are hard plastic, ventilated crates that stack. Most vegetables can be packed loose in them, then carried to the wash station and hosed off right in the crate or transferred to a tub for washing. They also make excellent storage and transport containers, and they are useful for organizing and storing tools and supplies elsewhere on the farm. They are so useful that it is worth expending some energy to locate a supply.

Bulb crates can often be purchased for a dollar or two each from wholesale greenhouses. Here's a tip for locating bulb crates: Go to a supermarket floral department in winter or spring and look at the potted tulips, daffodils, and hyacinths. You should find a tag with the grower's name, but, if not, ask the floral department employees

Bulb crates have dozens of uses around the farm, from harvest to storage. Used in the field for sweet potato harvest, they also can be stacked in the barn for curing and storage.

where they get their flowers. You'll likely find the name of a regional greenhouse grower that you can then look up on the Internet. These greenhouses buy thousands of bulbs for forcing, and those bulbs are almost always shipped from Holland in plastic crates. Some greenhouses have mountains of bulb crates that they are happy to get rid of; you may even be able to get them for free. Just be aware that they take up a lot of space, and you won't be able to transport more than a few dozen at a time, depending on the size of your vehicle.

Bus tubs are another low-cost option for moving produce; they can be purchased from restaurant supply companies. They are FDA-approved for holding food, and they are durable. Some can be purchased with lids, which are great for storing lettuce and other perishable greens in the cooler.

Wooden crates are often available free from supermarkets, which view them as a disposal problem. Produce shipped from afar often comes in very sturdy wooden crates that can be stacked. Certified-organic growers cannot use crates or boxes that contained

Ewell Culbertson of Pachamama Farm in Colorado hydrocools green beans by submerging the harvest crate in a vat of cold water.

conventional produce, but there are plenty of organic produce crates looking for homes, too.

Buckets are an essential harvest tool on market farms. You can purchase new 5-gallon buckets at hardware stores, but you can easily find them for free or at a nominal cost if you know where to look. Restaurants purchase a lot of supplies such as pickles and doughnut icing in 1-gallon, 2-gallon, and 5-gallon buckets. Check with supermarket bakeries and places like Dunkin' Donuts. Always get buckets that have been used for food rather than chemicals or industrial supplies, and clean them thoroughly.

When searching for harvest and storage containers, always look first at commercial suppliers, which are more likely to have sturdy containers that can withstand the heavy loads and banging around that they will undergo. Crates made specifically for produce harvest will last a long time. Look for crates that nest to minimize storage space. Don't buy cheap stacking crates from discount stores or office supply stores; they just won't hold up.

Food Service Suppliers

Restaurants and grocery store suppliers can be a great resource for market growers; they are in a similar business, after all. You can find all kinds of commercial kitchen equipment, storage containers, sign systems, packaging, and display baskets.

Most cities have one or more companies that specialize in resale of restaurant and supermarket equipment. Check the classified ads or Craigslist regularly if you're in the market for coolers, sinks, and other food service equipment. Many of these companies have regular auctions, and if you make it a habit to attend, you can pick up many useful tools at low prices. Even new equipment from a restaurant supplier can be less expensive than big-box store supplies, and usually more durable.

Although a local store is best, you can also buy many food service items by mail order from national distributors. Here are two great ones, whose catalogs you should acquire right away:

* Hubert supplies the supermarket segment of the food industry. The company's catalog lists everything from salad bar clamshells to sign systems to floral displays. www.hubert.com or 866-482-4357.
* US Foods offers culinary equipment and supplies. www.usfoodsculinaryequipment andsupplies.com or 866-636-2338.

⇒ THE PACKING SHED ⇐

Every farm needs a dedicated place for cleaning and packing produce. It needs to be clean, brightly lit, safe, and well organized. You will spend a lot of time in your packing shed, and it will become an important part of your food safety plan, so it bears careful consideration from the start. When choosing a location, think of the packing shed as the funnel from the fields to the vehicle that will transport your produce off the farm. It should be easily accessible when you are bringing in harvested produce, and you should be able to drive your vehicle right up to the door when you're getting ready to go to market.

The structure itself will be what you can afford when starting out. Some growers use a canopy to cover their wash tubs and other equipment. Some use a hoophouse or greenhouse. Others convert barns or build new buildings. Whatever the roof is over your head, here are some of the other things you need to consider:

* The packing shed needs clean running water that is safe for washing produce.

Healthy Farmers

For a time, the University of Wisconsin had a grant to help growers work more safely and ergonomically. The Healthy Farmers, Healthy Profits Project developed tip sheets explaining tools and procedures for preventing repetitive stress injuries and improving efficiency on the farm. Although the project is no longer active, the tip sheets fortunately are still available online. You will find great information about farm-built tools, such as lay-down work carts, rolling dibble markers, specialized harvest carts, a strap-on stool, hands-free wash station, packing shed layout, and much more. You can find the tip sheets at bse.wisc.edu/HFHP /tipveggy.htm.

* It should have food-contact surfaces that can be sanitized, such as hard plastic tubs or stainless steel sinks.
* Floors need to be of a material that is not slippery and can be cleaned, such as brushed concrete.
* Ergonomics have to be considered, so that you aren't straining any part of your body regularly while working. According to the Healthy Farmers, Healthy Profits Project at the University of Wisconsin, the most efficient work table height is halfway between wrist and elbow, measured when the arm is held down at the worker's side. For heavier items, it is slightly lower.
* Good lighting is essential so you can spot defects in your produce.
* The most efficient layout avoids extra steps and people crossing paths. It should move produce in the direction of the worker's leading hand—left to right for right-handed people.
* Packaging supplies should be kept nearby and neatly accessible.
* Rodents and other vermin, such as cockroaches, should be trapped regularly.

When growing on a small scale, most growers make do with hand cleaning of produce. Some produce (apples, tomatoes, or peppers, for example) can be cleaned with a dry cloth. Onions and garlic can be cleaned by brushing or peeling the top layer. For produce that needs to be washed, white plastic utility sinks or large Rubbermaid tubs are inexpensive and can be cleaned and disinfected easily with bleach. They also can be drained and refilled easily when water gets dirty. For food safety reasons, clean, potable water should be used for washing produce.

Produce can be dried in crates, on a frame of 2 × 4s covered with hardware cloth, or in plastic baskets. Some growers use plastic laundry baskets and spin them around to dry lightweight items. Many growers remove the agitator from a clothes washing machine, then put greens in it on the spin cycle to remove water. Those who grow a lot of lettuce or mesclun find that an electric or hand-cranked

A produce-washing station can be on a covered porch or other open-air structure, in a greenhouse, or inside a building. The important elements are excluding birds and rodents, using potable water, and having food-contact surfaces that can be sanitized easily.

Mesh Bags Save a Step

To make washing salad greens and other delicate produce easier, line your harvest containers with mesh bags. The bag can be lifted out of the container and put directly into the washing tub. Swishing the bag around will remove soil and hydrate the produce. The bag also can be placed in a salad spinner. Mesh bags in all sizes are available in the laundry section of discount, hardware, and home stores. They can be laundered and bleached to sanitize them between uses.

salad spinner, available from restaurant supply companies, is a labor-saving device. Whatever you use to dry produce, make sure it's as clean as the washing containers and sanitized before every use.

More specialized cleaning equipment is available for those who grow enough of certain crops to make them worth the cost. Barrel washers tumble root crops, such as carrots and beets, to get them clean. Conveyors with spray wands or brushes are available for many types of fruits and vegetables. This kind of equipment is widely available secondhand in fruit- and vegetable-growing areas of the country but hard to find elsewhere. Thanks to the Internet, virtually anything can be found online if you know what search terms to use.

❧ PACKAGING ❧

You will also need postharvest containers in which to store your produce once you have cleaned and graded it. We have found restaurant supply companies to be a good source of long-lasting containers for items that will be sold at farmers markets or through a CSA.

For produce that is going to a grocery store or restaurant, you may need to comply with

A barrel washer: a motor rotates the barrel slowly while nozzles in the pipe at top spray water on the root crops.
PHOTOGRAPH COURTESY OF JOSH VOLK

Salad Spinners Widely Available

Commercial salad spinners are available in 2½-gallon and 5-gallon sizes, with manual or electric spinners. Prices range from $120 for the smaller manual model to $600 for 5-gallon electric models to $2,000 for 20-gallon models. Local restaurant supply stores may have them, and you may find a used one locally or on eBay. You also can purchase one by mail from numerous suppliers. Shop around, because prices vary considerably for the same brand. The two food service suppliers listed previously carry them, and here are three others that sell the Dynamic salad spinner, pictured here, or a similar brand:

* www.restaurantequipment.com /GREENSMACHINE.html; 800-845-6677
* www.instawares.com; 800-892-3622
* www.akitchen.com; 888-388-9641

USDA's packaging standards, reprinted in the appendix. Find out first what the buyer expects, because there is a lot of variation. Some supermarkets will accept only new, standard packaging. Others don't care how you get it there as long as it's clean and counted or weighed correctly. You may be able to pick up used cartons at a supermarket and reuse them. If you're an organic grower, you should reuse only organic produce boxes; you will lose the organic label on produce that has been packed in a conventional produce box. Growers who get food safety certified have to provide tracing information on every box.

You'll find there are many variations in how produce can be delivered. For example, common packaging sizes for beets include 50-pound mesh bags; 45-pound wirebound crates bunched in 12s; 38-pound cartons bunched in 12s; 35-pound half crates, loose; 32-pound, ⅘ bushel crates; 25-pound bags, loose; and 20-pound cartons, bunched in 12s. Look at all the standards for all the different items you grow and try to find one that works for numerous types of produce; it will help you purchase boxes in larger quantities and therefore at lower prices. You'll find packaging suppliers listed at the end of the book.

❧ COOLING PRODUCE ❧

Coolers are a tremendous asset on a vegetable farm because they extend the shelf life of produce and allow you to harvest whenever the crop is ready rather than right before a market. Most produce items should be stored at 32°F to 36°F (0°C to 2°C) for longest storage life. Some things, however, will experience chilling injury at temperatures that low. Basil, snap beans, some melons, okra, peppers, and squash should be held at around 45°F (7°C). Cucumbers, eggplants, pumpkins, and watermelons should be kept at 50°F (10°C). Tomatoes should not be refrigerated at all.

In hot weather, field heat must be removed from produce as soon as possible after harvest to prevent it from spoiling. Some growers immerse it in tubs of cold water to bring down the temperature. It's important to use clean water for this, as recent research suggests that pathogens can be forced into plant cells by immersion. Washing and drying produce will also reduce its temperature by evaporative cooling. Some produce items can be iced immediately after harvest.

To get to market, growers use scale-appropriate cooling methods: iced containers or air-conditioned vehicles for small producers; refrigerated trucks for large producers. See chapter 4 for information on buying and building coolers.

Processing Kitchens

Faced with an overabundance of a crop, nearly every grower at some point contemplates the idea of processing. It's impossible not to look at strawberries (or tomatoes) destined for the compost pile and think, "This could be made into jams (or salsas) and sold year-round." And it's true that many small farms have successfully created processed foods to boost their offerings throughout a longer season. But processing is a completely different thing from farming, and you should do your research before you proceed.

First, you need to know what state and local regulations may apply to your venture. Every state has its own health rules, and local governments may have additional zoning or building codes that would apply to a processing kitchen. Most are based on the federal Food and Drug Administration's Good Manufacturing Practice. The description of Good Manufacturing Practice is available online and printed in the Code of Federal Regulations; if you want to read it yourself, ask a reference librarian to help you find 21 CFR, Part 110.

Basically, the Good Manufacturing Practice says to keep the place clean and make sure food can't get contaminated. It's a lot more detailed, of course. Some of the most costly requirements include:

- Walls, floors, and ceilings must be washable, and the kitchen must be ventilated so that drip or condensate from the ceiling or fixtures won't fall into food.
- Food-contact surfaces, tools, and equipment must be resistant to corrosion and made of nontoxic materials. Seams on surfaces must be smoothly bonded to prevent accumulation of food particles, dirt, etc.
- The room must be screened to keep out birds, insects, and other pests.
- You must have a bathroom if you have employees.
- You must have a hand-washing sink and separate sinks for washing, rinsing, and sanitizing equipment and utensils.
- Water must be from an approved source.

These guidelines leave room for interpretation. One grower reports that state health officials required him to put in a triple sink for cleaning utensils, a sink for food preparation, a hand-washing sink in the food preparation area, and a bathroom with toilet and sink. They put in tile floors, plastic wall covering, and stainless steel countertops and sinks. Most of the equipment came from auctions and restaurant suppliers. Growers report spending anywhere from a few thousand dollars to outfit an existing space to more than $100,000 to build a kitchen from the ground up.

Once you have an approved kitchen, you can make jams, jellies, and some baked goods without further ado. Because of their sugar content, they are not considered hazardous foods. Similarly, acidic foods, such as vinegars, can be made without special training.

For those foods that can pose a hazard, the FDA has strict rules. Processed vegetables fall into two categories based on the acidity of the product. A low-acid food has a pH higher than 4.6; an example would be canned green beans. An acidified food is a low-acid food to which sufficient acids or acid foods have been added to bring the pH to 4.6 or lower; an example would be pickles. The FDA considers salsa, sauce, and dressings to be acidified foods, unless they are held under refrigeration.

All foods in the acidified and low-acid categories can be made only under the supervision of someone who has successfully completed the FDA's Better Process Control School, which takes four days and is usually held at universities with food science departments. Furthermore, to make any foods in those categories, you must register with FDA as a processor, file scheduled processes for each product, maintain specific processing records, and use equipment that meets certain minimal standards. The equipment includes pressure canners.

If all the rules seem intimidating, rest assured that you are not the only farmer who has feared to venture into processing because of the regulations. There are many resources, though, to help you make the leap. Because many economic development groups consider value-added food products to be excellent business ventures, you may find help from your state's agriculture or commerce departments. Also, many state universities with food science programs will help entrepreneurs develop and test new products. Check first with your state Extension service to find out what resources are available. You may find that you can rent a processing kitchen nearby or even get funding to build your own. In a few places, there are groups of farmers who have built processing kitchens for members of their cooperative.

Our farm has some personal experience with processing. For five years, we were involved in a cooperative that processed tomatoes into salsa. The cooperative was structured so that farmers could be paid for their tomatoes and their labor in the kitchen in cash, or they could take product at wholesale prices as compensation. By selling the product at retail prices, the farmers could increase their revenue. We found two big flaws in the project: First, it was hard to get people to work in the processing kitchen during the peak season; second, small-scale production of "boutique" products requires a high retail price that might not be viable in some locations. There was definitely a steep learning curve and a lot of initial expense in developing a product line. My advice is to go into processing cautiously, aware that it's not the path to easy riches and that it will take dedication and a whole different set of skills from what you need as a farmer.

Growers who don't want to do their own processing have two options. First, they can find a co-packing arrangement in which a processor uses the farmer's produce but does all the work to produce a processed food. Second, some farms use what is known as "private label" processing, in which their name is used on the label, but the ingredients came from elsewhere. Private labeling is more common at farm stores and may be prohibited at farmers markets.

Pricing Your Products

Setting prices is one of the most difficult jobs facing the direct-market farmer. There is no single resource that can tell you what to charge. Yet it's essential that you know how much to ask for your products if you want to make a profit. In this section, I'll tell you about some of the pricing strategies other growers use and refer you to other resources on the subject. First, I'll address the question of pricing for retail, such as farmers markets and your own farm store; then I'll tackle the more difficult question of pricing for restaurants, grocery stores, and other wholesale accounts. But first, a few words about your cost of production, which should be your first consideration in setting prices for any market.

➤ COST OF PRODUCTION ◄

When you first start growing, you probably won't have a very clear idea of your cost of production for any specific crop. New growers tend to just lump all the crops together, add up their income, subtract their expenses, and then decide whether they made a profit or not. Sometimes you get lucky and do make money, but this is a risky way to approach the central issue of your business.

Instead, you must learn to track your expenses and assign them to each crop as either a direct cost (for example, the seeds for that crop) or as a percentage of overhead (the liability insurance on your farm). Refer back to chapter 5 for a discussion of enterprise budgets, which will help you figure out a system for determining your own costs of production. This number should include the salary you

want to pay yourself. The prices you set should use this as the base but also include a reasonable profit for the business.

If you are selling at farmers markets or your own farm store, you know that your prices must be in line with what other vendors are charging and what the supermarkets are charging. That does not mean you want your prices to be the lowest in town. In fact, you should probably be charging substantially more for your products if you are offering higher quality or some other perceived value. Always be careful that you are comparing apples to apples. A few years ago, for example, a grower in Oklahoma was upset when some customers complained that her prices were too high. She conducted a study in which she compared her certified-organic produce prices to prices charged at several supermarkets and a Wal-Mart store. In comparing 42 items, she found that, although unit prices were often lower at the supermarkets, her produce always weighed more per unit. As it turned out, her price per pound was often cheaper and always competitive with supermarket produce.

Unfortunately, many people believe that farmers markets are expensive, offering food just for wealthy people. Numerous studies have refuted this myth, but still it persists. So part of your pricing strategy must include an emphasis on value, while still getting the best price possible.

Grocery store prices are the starting point for setting prices. But there are several reasons why your prices might be higher than grocery stores:

* First, produce is often a "loss leader" for supermarkets. Produce managers buy big quantities of an item and sell it for barely above cost as a way of enticing shoppers into the store. Don't try to compete with these deeply discounted prices. You'll go broke.
* Freshness gives you an edge over supermarkets, and you should be able to charge accordingly. Potatoes, for example, may be selling for 50 cents a pound in the supermarket. But those potatoes may have been in storage for months, and their flavor will be far inferior to your freshly dug spuds. In fact, the two aren't even the same item. Freshness equates to better flavor in virtually all kinds of produce. Consumers know it and are willing to pay more for it.
* Quality should be higher at farmers markets than at supermarkets. Green beans that were mechanically harvested and have been sitting in a box for a week will be noticeably duller than your fresh-picked beans. Lettuce varieties that withstand a thousand miles of shipping will be tougher than your crisp greens. Charge more for higher quality.
* Specialty items just aren't available at many supermarkets, so you're going to have to make up your own prices. Heirloom tomato varieties, ripe and flavorful, are worth considerably more than the pale, hard tomatoes sold in stores. Purple potatoes, fennel bulbs, arugula, specialty melons—those kinds of exotic items command a higher price than some vaguely comparable supermarket item.

How do you know what is a fair price? It helps to know your costs of production. If you keep good records of your inputs and overhead, you can figure out how much you need to charge to make a profit. But you also have to take into consideration what other vendors are charging. Some markets have price guidelines that prohibit vendors from charging too little. Find out from your market manager if your market has any pricing rules. If not, survey the market regularly to find out what other vendors are charging. Find someone who attracts a lot of customers, who has high-quality produce and a good display, and use that farmer's prices as guidance.

➤ WHOLESALE PRICES ◄

Setting prices for restaurants and grocery stores is a much trickier business. There is no open venue where you can compare prices. Other vendors and buyers usually keep pricing information a secret. And that keeps growers off-balance because they never know if they are asking too much and will thereby lose the sale or asking too little and not making the margin they deserve.

So how do experienced growers determine how much to charge? Here are the most common options:

1. **Go with the market.** Find out what the wholesale distributor is charging for produce of a similar type and charge the same amount or slightly more for better quality. But how do you know what the wholesaler is charging? Your most important tool in this area is the US Department of Agriculture's Market News Service. The USDA collects and publishes pricing information for all kinds of produce on a weekly basis. Two of the most useful reports available online are the National Fruit and Vegetable Retail Report, which lists supermarket prices for commodities, and the National Fruit and Vegetable Organic Summary, which lists bulk pricing of organic produce at terminal markets. You can create your own custom report for conventional terminal market prices. Start at www.marketnews.usda.gov/ and choose "Fruits and Vegetables" to see what's available. Bear in mind that terminal market prices reflect what is paid by "first receivers," which can include distributors, so those prices may be marked up by 20 percent or more before being sold to restaurants and grocery stores.

The Problem with Low Prices

The biggest complaint farmers have about selling at a farmers market is that new growers or hobby growers often come into the market and sell their produce way too cheaply. Please don't be guilty of this. You'll be hurting other farmers, the market's image, and ultimately yourself.

There's a certain dynamic that occurs around market prices. At the base of this dynamic is the fact that people don't buy more food than they need. If someone wants a watermelon, they're going to buy one watermelon. They're not going to buy three watermelons just because the price is cheap. So low prices will not improve the market's overall sales. One vendor's low prices may shift the volume within the market, so that the low-price vendor picks up more customers. But a farmer can't keep artificially low prices indefinitely and stay in business. You have to make a profit. When you raise your prices later, you're going to alienate your new customers who think of you as the cheap place to buy. In the meantime, you may have hurt farmers who depend on the market for their livelihood. And you will have cheapened the tone of the market overall. Customers will start expecting bargains rather than quality, and the entire market suffers.

Many farmers have found that they sell out despite having the highest prices at the market. You know the old expression, "You get what you pay for." People believe that. If your prices are lower than everyone else's, many people are going to assume there's something wrong with your produce. Conversely, if your produce is of high quality and you charge more for it, customers will think you've got something special, and they'll want to buy it.

Most customers will respect you for charging a good price for your food; it shows you have pride in your produce. But occasionally you will have someone gripe about your prices. The best response is a polite one: "This is what I have to charge if I'm going to stay in business." Most people can understand that; some will get huffy and walk away, but who wants that kind of customer coming back every week anyway?

If you're in a state that charges sales tax on food, get a calculator and add the tax to the customer's bill. Many growers don't charge sales tax because they think it's too much trouble to add those percentages. But the farmer still has to pay the state at the end of the year. Charging sales tax is like giving yourself a raise of 8 percent (or whatever your sales tax rate is). And customers won't notice; they're used to paying sales tax.

It is far better to go to a new market with high-quality produce, charge what everyone else is charging, and build business slowly and honestly. You will win the respect of both farmers and customers. You can hold your head up at the end of the day. And, over time, you'll build a base of loyal customers who come to you because of the quality of your produce, and your marketing venture will be a success.

2. **Check prices at local grocery stores.** This is not a great way to find out the going price, because supermarkets and natural foods stores vary in the amount of their markup. Sometimes produce is much cheaper than you might expect because the store is using the low price to attract customers. Other times, particularly with the natural foods superstores, the markup is much higher than a regular supermarket. All those caveats aside, a general rule of thumb is that supermarkets mark up produce 130 to 140 percent. (Multiply your price by 1.3 to 1.4 to see what the expected retail price might be.) But the best way to find out current prices may be the way you would least expect:

3. **Ask the chef or produce buyer to tell you prices.** No kidding; many growers really do depend on their customers to tell them the going rate for their produce. This works only if you have a good relationship with the buyer and feel you can trust him or her to be straight with you. You might, of course, tactfully point out to the chef that your operation can remain financially viable only if you can get a fair price. Chefs who are happy with your quality and service have an interest in your success and will no doubt see the logic in helping you get your best price.

One strategy is to sit down with a chef before the growing season and ask what he or she would have to pay for a particular item from a wholesaler. If you think the price seems reasonable and you can produce the item for that price, offer to grow it for the chef and keep your price constant throughout the season. That's a great help to the chef, who can put the item on the menu and know exactly how much it's going to cost. The same is true of grocery stores; buyers like a consistent price because it helps with planning. If you are sure you can make a profit from a certain price, everyone wins.

Still, it wouldn't hurt to have access to current prices from a market report just as a backup. Keep in mind that produce is a tiny percentage of chefs' total costs. Substantially higher prices to you will affect their per-plate price very little. Chefs who try to cut corners by shorting you are headed for trouble anyway, and you should be looking for new places to sell.

For some specialty produce, you are not going to find wholesale prices from Market News because the quantities are insufficient for them to be included on the report. When that happens, you can ask the buyer. But most likely, you're going to have to depend on the fourth method of setting prices:

4. **Determine your costs of production, and set prices to make a profit.** This is the most logical yet least common method of setting prices. Some growers who do it say their prices often come in much different from the wholesale price—sometimes more, sometimes less. Chefs get accustomed to working on a different system and, if they're happy with the produce, will not complain.

Still, keeping track of production costs is a difficult thing for most new growers to do because it involves keeping records about all inputs and all sales and then analyzing the data to determine where prices should be. It's a headache in the midst of a busy season to keep count of every beet you pull or tomato you throw in the compost pile. But those kinds of records will be invaluable to you in the future to help you decide what to grow, as well as how much to charge, so you should start your business on the right foot by keeping good records. (I know I have said this frequently throughout this book, but I don't think you will find a veteran grower who would disagree. Good records = success!)

Farmers Market Presentation

Because most growers get their start selling at a farmers market, I want to address that topic in detail. Many of the comments about presenting produce also apply to other markets, such as CSAs and farm stands.

Over the years, *Growing for Market* has published numerous articles about the art and science of farmers market selling. The comments below have been culled from several of these articles; some of the people quoted may now be in different places or positions, but their advice still stands. You'll hear from Barry Benepe and Tony Mannetta, former administrators of New York City's Greenmarkets; Mark Phillips, who grows 125 acres of vegetables near Milford, New Jersey, and sells at the Greenmarket; and Marion Kalb, director of the Southland Farmers Market Association in Los Angeles.

One of the maxims of selling anything in a retail setting is that you have about three seconds to catch the customer's eye. At a

Recipes Sell Produce

When customers are inspired to cook with fresh food, they buy more. Go to market prepared to inspire them with recipe cards for whatever items you are offering. I have commissioned a food writer to create a book of reproducible recipe cards featuring recipes that use a lot of produce. That book, *Farm-Fresh Recipes*, is now available as a downloadable PDF. It contains almost 300 recipes for fresh produce, laid out so that you can run off copies on your printer and cut them into recipe cards to give to customers. It's free for *Growing for Market* subscribers, or it can be purchased in the online store at www.growingformarket.com.

farmers market, the window of opportunity may be even smaller, with so much color, fragrance, and activity calling out to shoppers from every direction. With so little time to make your sale, you need a display that will stop shoppers in their tracks and pull them to your market stand. And how do you do it? First, learn the tricks of the trade—merchandising strategies that work for supermarkets and that will work for you, too. Go to the supermarket and survey the produce section. Notice what catches your attention as you scan the displays, and try to figure out why that particular display appeals to you.

"You really have to think like a retailer, because that's what you are—you're putting a store in the back of your truck," says Mark Phillips. "Any farmer who just puts stuff on tables and hopes it sells is probably going to be disappointed."

❧ APPEARANCE OF BOUNTY ❦

Barry Benepe recalled the grower who displayed green beans in a small bowl, which she refilled from a basket in her truck. The effect was that no one saw her green beans. In contrast, another grower made a mountain of radishes and the splash of color drew people from across the street. "Quantity makes a big difference," Benepe says. "If you have it, flaunt it."

Phillips agrees. "I won't put one crate of lettuce out; I'll put 10 out and bring a half of one back home and throw it away."

Mannetta recommends that you stack your produce to create an appearance of bounty and to make it more visible from a

The old saw "Stack it high and watch it fly" works at farmers markets. This vendor piled up his bags of red, white, and blue potatoes as high as they would go—and then added another pile beside it for extra visual punch.

Tilting crates toward the customer creates an inviting display.

distance. Start stacking from a few feet above the ground (never put food on the ground), and make tiers that reach above your waist on the table. Every item should be tilted to give the customer a better view and make your supply look larger. On the tabletop, don't settle for just one level of produce. Stack boxes at the ends and stretch a board across them to create a second tier for smaller items.

Even if you don't have large quantities of produce, you can create an illusion of bounty by turning baskets on their sides with the produce spilling out from them.

≫ Using Color ≪

Mixing colors is an important aspect of the art of display. Group your produce to create large blocks of color, then group produce of a contrasting color beside it.

"Yellow is a real eye-catcher," Mannetta says. "Farmers use it on the corner of a display, even if they have only a little bit."

Phillips likes to make checkerboards of black and red raspberries or yellow and red tomatoes in quart boxes, tiered on red plastic crates. Each year Greenmarket gives awards to vendors for creative displays. One grower received mention for his "waterfall" of potatoes, an inclined board covered with spuds of all shapes and colors. Another won for using wicker flower-gathering baskets to display her colorful collection of peppers, eggplants, and squash. Another lined up his buckets of flowers according to colors of the spectrum, creating a rainbow effect from a distance.

Kalb says that one of the standouts at the 17 markets she directs in Southern California is a citrus grower who uses a signature color,

Mixing up pints of berries and tomatoes creates a colorful quilt effect that catches the eye.

Beets keep their intense colors when they're misted throughout the market. Always put the colorful side of your produce toward the customer.

Plastic may be practical, but natural baskets and wooden crates have a warmer look, especially when you don't have a huge amount of an item to display.

Many vegetable farmers grow a few rows of sunflowers just so they'll have something bold and gold to display at market. Yellow is the most visible color, so put it on the outside of your booth.

These growers have all their produce prepacked so they don't have to bring a scale or calculate fractional prices. Notice how the asparagus bunches are standing up.

orange, every place he can: his sign, his cap, his apron, and his table. Flags and banners are a popular technique, too. A strawberry grower flies a row of banners with strawberries on them. Colorful flags, medieval-style banners, and windsocks are now widely available in gift and garden stores, or they can be purchased from a specialty flag company.

Go easy on the plastics, though, Mannetta advises. "We really feel that wood is the way to go," he says. "People are looking for that natural feeling, and wood brings home that message."

The neutral colors of natural materials help accentuate the bright colors of the produce. Baskets are a perfect display medium, and even cardboard boxes can be wrapped in burlap sacks to look more "natural." One grower makes "nests" for large items of produce by wrapping sisal rope around the crates.

Mannetta says that a dozen buckets of flowers on the corner of a stand will catch glances from far away and attract customers like a magnet.

Also, don't let your produce get dry and dull-looking. Mist it regularly to keep the colors bright.

In some cases, spreading the produce out on a table creates a greater appearance of bounty.

Samples at Market

Farmers markets are one of the best places for educating consumers about the benefits of buying locally grown produce. And one of the best ways to educate people is to let them taste for themselves. One taste of a truly vine-ripened heirloom tomato will sell them on several pounds. A whiff of a ripe melon will have them lining up to make a purchase. And nothing sells value-added products better than a taste of a recipe using the product.

But offering food samples at an outdoor farmers market carries a food safety risk. Some states and local health departments may have rules pertaining to sampling, so ask your market manager for guidance. If there are no rules, here are some basic food safety guidelines to make your samples appetizing and safe:

1. Always use food service gloves when handling samples, and replace them after you touch something else, especially money.
2. Wash your hands thoroughly before putting on your gloves.
3. If possible, bring samples already washed and cut in food safe containers on ice. Use toothpicks or individual cups for serving. Prepared samples must be protected from droplet contamination, insects, dust, and a customer coming in contact with more than their own sample. All samples should be stored in food-grade, nonabsorbent (not wood) containers.
4. If you have to cut samples at market, knives should be stored in sanitizer solution when not in use, and cutting boards and containers must be sanitized and air-dried before you use them. The three acceptable sanitizing solutions, which can be used as a dip or a spray, are:

- 100 ppm* chlorine—½ ounce (1 tablespoon) per gallon of water;
- 200 ppm quaternary ammonium—½ ounce per gallon of water;
- 25 ppm iodine—½ ounce per 2½ gallons of water.

These solutions are good for four hours at your market stand.

---

* "ppm" stands for parts per million

➤ SIGNS ◄

People don't want to have to ask for prices, and they don't want to appear ignorant, so label everything, these experts say. Start with a sign full of information about your farm. "We grow everything we sell," one farm reassures customers, even though that's true for all the other vendors, too, since Greenmarket prohibits people from bringing in produce they didn't grow. Single-commodity farmers, particularly those who can't put their product on display, need bright, eye-catching signs. A big black-and-white cow advertises a dairy that has to keep its products refrigerated. A bright yellow chicken does the job for a poultry farmer.

Customers should not have to ask the name or price of anything. This vendor has clear price signs on every item.

Signs should provide the information to entice your customers to buy. "Rocambole garlic—large cloves, easy to peel" is much better than just a simple "Garlic" sign. Some growers paint the name of the item on wood and write the price on a card that can be attached, giving them flexibility to change prices.

Kalb recommends hanging a laminated fact sheet about yourself, your farm, your community, or your farming practices. If you're a member of the school board or a blue-ribbon winner for your pecan pies, put it on your sign. The idea is to humanize your business transaction, to establish a relationship between you and the customer that goes beyond taking their money.

Price signs are so important that some markets require them on all products. Some growers say they don't post prices so that customers will be forced to come up to the stand and open a conversation. Kalb says that is a bad idea. "I think signs make people more willing to come to your stand," she says. "Many people may want to come over and see what you have, but they don't want to ask a question." Price signs also assure customers that everyone is paying the same price.

Market customers love to be educated about the food they're buying. This grower created signs to tell the story of each heirloom tomato variety.

Suggesting ways to use produce helps sell more.

Signs can be used to tell customers a little bit more about you, your farm, and your practices.

Market Shade

Although some markets require growers to have a particular type of canopy, the Fayetteville, Arkansas, market welcomes individuality. Artist Kent Landrum captures the colorful scene on a Saturday morning.

Most farmers market vendors use canopies or umbrellas to shade their produce and define their space. Some markets require the same kind of canopy for all vendors, to provide a uniform look. Other markets prefer a colorful mix of umbrellas and canopies. The most common is the pop-up canopy, which is quick and easy to erect. Brands include E-Z UP and Caravan. Although they are often available at discount stores for a couple hundred dollars, those are usually lightweight models not intended for the hard use of a weekly farmers market. You'd be smarter to spend a few hundred more for a heavy-duty commercial model. Ask other vendors at market where they purchased their canopies,

or check the resources at the end of the book for mail-order sources.

The biggest issue with using tents is keeping them anchored in wind. Any kind of canopy, but especially tents with side panels, can become airborne if not properly weighted down. Wind-blown objects are the leading cause of injury at farmers markets. Some markets require vendors to have 24 or more pounds of weight on each tent leg. Several suppliers sell bags that you can fill with 40 pounds of sand and attach to the legs.

Many growers have come up with their own systems for holding down their tents. The most common system is to fill a paint can or bucket with concrete and set an eye

Whispering Cedars Farm uses two pop-up tents to protect a huge display of cut flowers at the Lawrence, Kansas, market.

bolt into it as it hardens. The container can be painted for a neater look if necessary. Bungee cords are recommended for connecting the weight to the holes in the tent legs because they flex a little in a breeze. Other people use buckets full of water tied to the tent, but some markets don't allow this because the handles can break. Some use old window weights, weight machine weights, and other castoffs found around the farm and at garage sales. In really strong winds, growers often lash their tents to their vehicles or nearby poles.

Whatever system you devise, be sure it doesn't present a tripping hazard to you or your customers.

A canopy blown by the wind can cause serious injury to customers and vendors, so most farmers markets require vendors to secure their canopies with weights, such as these commercially available weight bags.

At Greenmarket Keith Stewart won an award for his sign technique: He chiseled a slot in the wooden crates he uses for display, then cut plywood squares to fit into the slot. Each square has the name and price of an item painted on it (prices don't vary much over the season at Greenmarket), so he can shuffle through his deck of price cards and quickly insert them into the proper crate.

❧ ATTRACTING ATTENTION ❧

The experts stress that you shouldn't display produce on the ground, where it can be kicked, marked by dogs, covered with dust from traffic, and thrown into the shadows. Tables at elbow to eye level are ideal. Kalb recommends a table with enough room for customers to lay down purses and bags so they can comfortably help themselves to your merchandise. Bales of straw are a great substitute for a table, particularly with fall crops, and sometimes you can even sell the bales, too.

Phillips advises you to use fragrance to attract customers. Squash a couple of strawberries, and the smell will waft through the air. Brush the basil every time you walk past it. Phillips suggests that you put as much stuff between the customers and the checkout as you can, so that they will have an additional opportunity to buy more on their way to the scales.

Mannetta suggests bringing things from the farm to attract attention. A cider press, even if it's not being used, brings over curious shoppers. A pedal-powered bean sheller is popular at a market in Georgia. Pepper roasters entice shoppers with their fragrance. If allowed,

A vendor roasts peppers at a farmers market.

animals can't be beat as an attention-getter. Every parent in the market will bring their children over to pet your lambs or look at your rabbits.

"A little whimsy never hurts," he adds. A wheatgrass grower covers his truck with flats of wheatgrass, making it appear to be covered in a carpet of green. Another grower advertises "The Official New York Mets Souvenir Cabbage."

Finally, Mannetta says, "No display is complete without a smile." An artistic display may bring the customers over to your stand, but it's your attitude that is going to make the sale. A friendly, warm demeanor will keep them coming back week after week.

If you need to charge sales tax for your products, either add it at the register or include it in the sales price—and let your customers know it's included.

Shelves make the most of limited space and put products closer to customers' eye level.

Credit Cards and EBT

Farmers have benefited enormously from recent developments that allow them to accept credit cards, debit cards, and Electronic Benefit Transfer (EBT) cards for food stamps and other food assistance programs. Using smartphones, wireless terminals, iPods, and tablets, farmers and farmers markets are now able to take most forms of payment, even when selling at markets without phone lines or electricity.

At many farmers markets accepting EBT and credit cards, the market manager has a wireless terminal where customers can swipe their cards to purchase tokens or scrip to spend at the market. But the time is fast approaching when all vendors will be able to accept EBT cards at their own booths. It's already possible for any farmer with a smartphone and mobile payment application to accept credit and debit cards. And the federal government has been providing grant funding to get food assistance programs, such as the Supplemental Nutrition Assistance Program (SNAP) and Women, Infants, and Children (WIC) program into farmers markets.

This is a rapidly changing environment, so rather than providing a list of companies that offer the service, I suggest you do an Internet search for "mobile payment applications" and see what is currently being offered. Compare all fees and discount rates to find the best deal for your business.

A colorful banner makes a market stand memorable. Hanging it on the canopy means customers won't block it when standing in front of the display.

CHAPTER 8

Managing Your Business

Many market gardeners get into business by gradually transitioning from home gardening to commercial gardening. If you are somewhere on that continuum, stop right now and decide that from this day forward you will act like you're in business. Being a business owner means you have to pay attention to a lot of things that have nothing to do with growing plants. But trust me: You've got to do it if you want to keep growing those plants.

In this chapter, I'm going to tell you about some business management issues that have proven to be very important to market farmers. These topics include taxes, legal structure, hiring, and insurance. I want to make it clear from the outset that I'm a market farmer and a journalist. I'm not an attorney, tax accountant, Internal Revenue Service (IRS) auditor, or insurance agent. As a result, I am not qualified to give you the absolute, up-to-date, definitive answers to specific questions on those topics. What I am qualified to do is to raise your awareness of these issues—and advise you to get professional help with the details of how these things affect your business. Let me say that again: Get professional help, especially in the beginning. Talk to small-business and farm-business experts in your state government to find out what you need to do before you sell your first tomato. Pay an attorney if you want to set up your business as a corporation. Find a good tax accountant who does farmers' tax returns, and have him or her do your tax returns for a year or two. Find an insurance agent you trust. I have been impressed with the services offered by the Small Business Development Center in my state. The center offers workshops on numerous topics you'll need to learn, and you can request a consultation with a small-business expert for guidance in some of the start-up issues you will face.

Tax Issues

The first thing you should do as a farm-business owner is to delete the word "hobby" from your vocabulary. Never call your farming business a hobby or your farm a hobby farm. If anyone teases you about your hobby, correct them immediately. This is not a hobby; it's a business. Be very clear.

I stress this point up front because the worst thing that can happen to your business, financially, is to have it declared a hobby farm by the IRS. IRS rules provide generous benefits for agriculture, but they are stingy about who qualifies as a farmer. If you don't treat your business like a moneymaking enterprise, you may not qualify for the deductions that make farming profitable.

The IRS has numerous standards you should meet in order to consider your farming venture a business rather than a hobby. "A hobby is an activity for which you do not expect to make a profit," says the IRS article entitled "Is It a Business or a Hobby?" In determining whether you are carrying on an activity for profit, the IRS takes all the facts into account. No one factor alone is decisive. Among the factors to consider are whether:

1. You carry on the activity in a businesslike manner.
2. The time and effort you put into the activity indicate you intend to make it profitable.
3. You depend on income from the activity for your livelihood.
4. Your losses are due to circumstances beyond your control (or are normal in the start-up phase of your type of business).
5. You change your methods of operation in an attempt to improve profitability.
6. You or your advisors have the knowledge needed to carry on the activity as a successful business.
7. You were successful in making a profit from similar activities in the past.
8. The activity makes a profit in some years.
9. You can expect to make a future profit from the appreciation of the assets used in the activity.

An IRS Audit

Donna Buono of Morning Song Farm in Rainbow, California, purchased her sub-tropical fruit and nut farm in 2001 and set to work reviving the neglected place. Two years into the arduous task, she was hit with an IRS audit that took up months of time and $2,000 worth of accounting fees.

"The audit revealed little amiss with my financial records," Donna said. "So we were pretty blown away with the IRS's next move. They said they were considering disallowing the entire farm, labeling it as a hobby instead of a business. If the IRS were successful in disallowing the entire farm as a business, we would owe considerable back taxes for 2001 and would be unable to count as business expenses any of the farm's expenses for 2002. We would have to immediately put the farm up for sale, just to cover the IRS bill!"

The IRS told Donna that it could classify her farm as a hobby farm unless she could produce a detailed, verifiable accounting of every hour she and her husband had worked on the farm in 2001. The labor needed to amount to at least 500 hours for herself and her husband—employee time did not count.

"That just amazed me," Donna said. "I had a grove manager full-time in 2001, and another $20,000 in day laborers' salaries."

By the time Morning Song Farm was being considered a hobby by the IRS, Donna and her husband had cleared 5 acres, installed microirrigation, planted 500 new avocado trees, regraded roads, automated the irrigation system, become active in farmers markets, and completed their organic certification. They had joined numerous business associations and purchased hundreds of rare fruit trees. They'd installed electric fencing and added livestock. They'd built a website that included online ordering for their macadamia nuts.

"We sure thought we were in business," Donna said.

The IRS sent the Buonos a form telling them to make additional copies for each day they worked on the farm in 2001, what work they had done, how much time it took, and how the task might be verified. Fortunately, Donna had kept extensive records: schedule calendars, a detailed daily farm journal, copies of all letters sent, transcripts of phone calls, and every email relating to the farm.

"It still took me 140 hours to compile the records for the first six months into one concise manuscript as they required. They'd demanded 12 months of data, but after documenting the first half for the year, I'd proved we'd not only met the 500-hour requirement, we'd substantially exceeded 1,000 hours."

Donna warns that beginning farmers need to know how much the IRS can demand of them, and that personal journals may even be subpoenaed.

"You might think you could just make something up in the event of an audit, but fabrication takes a lot more time than transcription; you won't be given a lot of time. Even throwing myself into it as I did, and having excellent records to work from, I got through only half the year before the deadline was upon me. I had no idea the IRS counted time. My journal saved me."

Note that you aren't expected to meet every one of these standards. These are simply factors that you should be aware of from the outset, so that you can conduct yourself accordingly. You never know when you might get audited, so it's smart to know what the IRS is looking for when conducting an audit.

You will never have to worry about proving you're not a hobby farmer if you meet the standard that the IRS calls presumption of profit: For most types of farms, your business is presumed to be carried on for profit if it produced a profit in at least three of the last five years. A profit is simply that you show more income than deductions on your Schedule F. In other words, if you show a profit three out of five years, you are safe—the IRS will not question your farming as a legitimate business activity.

The tax laws allow some leeway in determining how much of your expenses to deduct each year, so you do have a measure of control over whether your farm shows a profit or not in some years. With some expenses, for example, you may have a choice between depreciating them over several years or deducting the entire cost in a single year. You can choose which method to use based on whether you need to show a profit that year to meet the three-years-in-five standard.

Even if you don't show a profit that often, you are not automatically considered a hobby farmer. If the IRS should question you, the factors listed in the previous section will all figure in the IRS determination about whether your business is businesslike enough. Having the IRS audit your tax returns, however, can be a grueling process, as the sidebar on Morning Song Farm shows.

❧ RECORD-KEEPING ❧

The first rule of starting a farming business is this: Keep good records. From day one, keep a file of farm-related expenses. The money you spent on this book, for example, should be recorded and the receipt filed. If you buy other books or magazines while researching the possibilities, keep your receipts. Also, keep track of your time in a journal or day book. Write down everything you do pertaining to starting your farming business. If you spend an hour reading this book, write it down. Get in the habit now of recording your work-related time and expenses, and it will soon become second nature to you.

The tax benefits of farming begin in the year you file a Schedule F form with your income taxes. The Schedule F is the federal tax form where you report your farming expenses and income. If your farm shows a profit, that number will be transferred to your 1040, and you will pay taxes on it. If your farm shows a loss, that number will likewise be transferred to your 1040, and it will reduce your overall income, thus reducing your taxes. So obviously it makes sense to declare a loss on your farm if you have other income from another job, investments, and so on, right?

Not necessarily. The IRS is going to scrutinize your Schedule F to see if you are a serious farmer (rather than a hobby farmer). So the first year you file a Schedule F will be the year that you have some farm income. You don't have to make a profit that first year, but you should make it clear that you are a serious, start-up farming business.

➤ DEDUCTIONS ◄

The list of things that can be deducted from your taxes is a long one for farmers. IRS Publication 225, "Farmer's Tax Guide," goes into great detail about what you can deduct and when you can deduct it. It is available on the web at www.irs.gov. Your accountant also should be well versed in allowable deductions for farming (and if he or she is not, you should find a new accountant with experience doing farm tax returns).

In general, all operating expenses for your farm are deductible in the year you spend the money on them. That includes all your seeds, plants, supplies, small tools, employee wages, and real estate and personal property taxes on your farm assets. You may also deduct mileage at the standard mileage rate for any business-related miles you drive in your cars and trucks.

You may also deduct a portion of other expenses that are used for both your business and your family. These include a portion of rent, water, heating oil, telephone, repairs, insurance, interest, and taxes. It's not always easy to separate business from personal expenses on these items. The IRS says that any reasonable allocation is acceptable, based on the circumstances in each case.

For a business you are just planning, it's a good idea to track all those kinds of expenses the year before you start up so you have some baseline information. After you start your business, then, you will be able to determine how much expense is due to the new business. For example, keep a record of your water and electricity use the year before you buy a greenhouse. Then, once your greenhouse is up and running, you'll be able to figure out what portion of the electric and water bills are farm expenses. Depending on your infrastructure, you may be able to have separate meters for farm use of some of your utilities. If you must intermingle your family and farm utility usage, you may be able to borrow a power measurement tool from your electric company or purchase one. For example, if you use an electric heater intermittently in fall when you're washing produce in the packing shed, a power measurement tool will help you determine how much that heater is costing you.

Technically, all the expenses you deduct must be genuine costs of running your business. But when your business is also an activity that you enjoy, your pleasures become tax-deductible. What country dweller wouldn't want a new tractor with a front-end loader? When you're in the business of farming, you can buy one without qualms—it's all tax-deductible! Or maybe you would like to grow 50 varieties of lilies in your garden just because you love the look and fragrance of lilies. When you are in the cut flower business, you can grow all the lilies you want and deduct the cost. Are gardening books your passion? They're tax-deductible. How about travel to conferences or farm tours? Deductible. When you make your passion into a business, you can write off all kinds of things, effectively reducing their cost by 25 to 28 percent, depending on your tax bracket.

Building equity is another financial benefit of farming that many people fail to understand. Farmers are constantly improving their properties with barns, fences, trees and shrubs, graveled roads, and so on. Over the years, these improvements can add much more value than the cost of the improvements themselves. They can transform a rural house into a country estate. When the time comes to sell your property, you may find that your farm is worth much more money than you would have gotten from simple appreciation in real estate values. Your property is much more desirable with the features that you have added. And you were able to enjoy them all the years you lived there!

➣ PROPERTY TAXES ➤

Income taxes are not the only tax burden that is lightened by operating a farm business on your land. In nearly every state, farmland is taxed at a lower rate than residential or development land. In three states—Wisconsin, Michigan, and New York—farmers can claim state income tax breaks to offset local property taxes.

Farmland Assessments

Virtually everyone involved in farming agrees that farmland should be taxed at lower rates than other types of land. Without farmland assessments, most farmers couldn't afford to pay property taxes and would have to quit farming. The level of farming activity required to get a farmland exemption varies from state to state. For example, in New York farmers need to earn $10,000 on 7 acres to get the farmland assessment; other states require only $1,000 in annual sales.

The issue seems to rear its head every few years, when some enterprising reporter discovers a celebrity who pays low taxes by claiming to be farming their land. In 2012, *The New York Times* revealed the names of several New Jersey residents who had farmland assessments on their estates, saving them tens of thousands of dollars in property taxes. Among them were rock stars Jon Bon Jovi, who raises honeybees, and Bruce Springsteen, who leases part of his land to an organic farmer; the publisher Malcolm Forbes, Jr., who raises cattle; and Jon Runyan, the former NFL star and Republican congressman, who raises miniature donkeys and sells firewood off his estate. New Jersey officials said they were considering tightening the requirements for the farmland assessment from $500 in sales to $1,000 in sales.

By way of example, I asked my county appraiser to tell me what would happen if I stopped farming on my 20-acre property. He ran the numbers and informed me that our property taxes would increase 70 percent.

Farmland tax breaks have been on the books since the 1960s, a policy response to the destruction of valuable farmland. On average, more than 1.2 million acres of land are converted from agricultural use every year in the United States. That's about 2 acres lost every minute of every day! The reasons for farmland tax breaks are threefold: They help farmers survive by reducing taxes; they assess farmland appropriately, based on the smaller amount of income a farm will earn as compared to another use (such as a residential development); and they give farmers a financial incentive not to sell their land for development.

The rules for getting land assessed as farmland vary from state to state but are enforced by the county or town assessor where the land is located. If you are unsure about how your land is assessed, contact the office that sends your property tax bill. Inquire about farmland assessment to find out what rules apply and how you can get your land designated as farmland for tax purposes. In most cases, you will need to show the assessor your Schedule F to prove you are farming as a business. Again, a hobby farm will not qualify for farmland tax breaks.

❧ STATE REGISTRATION ❧
AND LICENSING

Once you're ready to start your farm business, register with the state. Most states have a publication that will tell you what is required to officially be in business, so check with the state department of commerce or revenue. It's hard to make blanket statements about state requirements, because there is a lot of variation among the states. In general, though, registering your business will trigger some reporting and new taxes. But it will also give you a tax number that you can use to purchase from wholesalers and make some purchases tax-free. Many companies in the horticulture industry will not sell to you without this tax number, so it behooves you to get it when you are ready to spend on plant material and supplies.

Some states also require that you collect sales taxes for farm products, even food. You will be required to remit those taxes to state government on a regular basis; the frequency will be determined by how much you sell but could be as often as every month.

If you have employees, your state registration will also trigger state employment taxes, including possibly unemployment taxes and workers' compensation insurance, though this varies from state to state, with agriculture being exempt from both in some states.

❧ LEGAL STRUCTURE ❧

Businesses, including farms, have a choice of legal structures under which to operate. Most farmers choose the sole proprietorship, which doesn't require much in the way of setup or maintenance. As a sole proprietor (which can include both you and your spouse), you *are* the business—you're responsible for all debts, taxed at the individual rate, and you are personally liable in the case of legal action.

Other legal entities may be preferable as a way of limiting your personal liability. These entities can include Limited Liability Corporations, subchapter C or subchapter S corporations, general partnerships, and limited partnerships.

The legal structure you choose for your business can influence many factors, including tax planning, liability, estate management, and passing the business to a new generation. It is far beyond the scope of this book to try to provide any advice about what is best for your situation. You can find a good publication explaining the options from Kansas State University; the link is listed in the resources. Even then, you would be wise to consult an attorney or accountant before deciding on the best structure for your business.

Hiring Help

A rule of thumb about hiring employees in a market farming business is that one person working full-time can handle only 1 acre of production. A couple, therefore, can handle only 2 acres of production. That is a gross generalization, and there are many exceptions in both directions, but it's a helpful starting point.

Most farmers quickly find out that they need to hire help if they're going to grow enough produce to make an adequate livelihood. Unfortunately, hiring help is a huge psychological barrier for many people. They may be afraid of the red tape associated with payroll and taxes; they may be afraid of hiring the wrong people and then not being able to get rid of them; they may worry about having to learn the labor laws as they pertain to interviewing, hiring, and firing.

Certainly, hiring employees for the first time is a big step. Good workers are hard to find, and it's easy for a farmer to unwittingly violate some federal, state, or local law. That's why it's so important for a beginning farmer to get professional advice, read the regula tions pertaining to employees, participate in grower associations, attend conferences, and seek advice from Extension specialists and veteran growers about becoming an employer.

Check with a Small Business Development Center (SBDC) in your state to see if there are classes or publications that will help you get acquainted with the labor laws, payroll taxes, and so on. You may find an SBDC advisor who will help you learn the requirements, or you can pay for a session with a farm tax accountant who will advise you. The IRS will provide you with either a paper or online copy of the *Agricultural Employer's Tax Guide*. It's a big publication, because it covers farms of all types and sizes, but it's written clearly enough that you should be able to go through it quickly and get a grasp of what portions pertain to your farm. If you are a beginning farmer and hiring only one or two people, your tax reporting and depositing requirements are not too bad—for most small-scale employers, taxes can be withheld from employees' paychecks and paid at the end of the year.

➤ DOING PAYROLL ◄

If you are not comfortable doing your own payroll, you can outsource the job. Check around for a local payroll service or bookkeeper who can do it. If you want to do the work yourself, with some guidance, there are many online payroll services, such as QuickBooks Payroll, that will calculate payroll taxes and tell you how much to write on the paychecks and how much to deposit into your tax accounts. Some of these services, however, do not support Form 943, the agricultural employer's reporting form. That should be the first question you ask before deciding whether to sign up for a payroll service.

You can even hire payroll companies to write the checks and make the tax deposits for you (which they do by debiting your bank account). I use one of these services, ADP, because it seems like an efficient way to outsource some of my administrative work. Once a month (you can choose any frequency, but once a month saves money and seems to be something most people can live with), a representative of the payroll service calls me to remind me it's time to do payroll. I tell her the number of hours each person worked. The next day, FedEx brings a package that includes paychecks for each employee, which I only have to sign, and a memo of the amount that has been debited from my account. I enter the amounts of the paycheck, the tax deposit, and ADP's fee into my checking account program, and payroll is done. The payroll company pays the taxes and files all the required reports with the federal and state government. At the end of the year, the payroll company creates W-2s for each employee and again files the required tax forms. It all costs me about $700 a year—a bargain, in my book. I could also choose to enter hours in an online program or to have pay direct-deposited to my employees' checking accounts.

If you are the kind of person, like me, who hates to do paperwork and can't keep track of deadlines for filing tax forms, get a payroll service, at least initially. Your mind will be put at ease, you'll have a model of how to do it in the future if you decide to take over the job yourself, and you can spend your time worrying about more interesting things.

❧ WHERE TO FIND GOOD WORKERS ❦

There are basically four different pools of people who could become your employees. Here is some advice about each:

Your children

Hiring your children can be a great deal for your entire family. Not only will you teach them skills and responsibility, you will also create some healthy tax benefits. The money your kids earn can reduce your own taxes and can be used as a tax-free vehicle for college or other savings. Here's how it works:

When you hire your children under age 18 to work in your business, you do not have to pay Social Security, Medicare, unemployment insurance, or other payroll taxes on their wages. Note that this applies only if your farming business is legally established as a sole proprietorship or a partnership in which both partners are the child's parents. If the farm is set up as a corporation, an estate, or a partnership involving someone who is not the child's parent, then

The author's kids worked on the farm throughout their teen years, then grew cherry tomatoes on their own in the summers to help pay for college.

the child's wages are subject to all the usual taxes paid by other, nonfamily employees.

Your child can earn up to the standard deduction—$5,950 in 2012—without paying income tax, either. And if your child earns more than $5,950, the tax rate will be 15 percent rather than the higher rate you would pay on the income if you didn't reduce your income by paying your child for working.

To avoid problems with the IRS when hiring your children, remember that the work they perform must be necessary to the business—work that you might reasonably pay someone else to do. However, the chores must reflect the child's age and ability; you might pay your 17-year-old to manage the farmers market stand alone, but not your eight-year-old. The amount you pay your children must also be reasonable and consistent with what you would pay another worker—you can't pay your kids $15 an hour if you're paying other workers minimum wage. Your children should keep track of their hours on a timesheet, and you should pay them regularly—with a check, so you have a paper trail. To pay them in a lump sum at the end of the season might arouse suspicion in case of an IRS audit. Keep good records showing what kind of work your kids do, the hours they work, and the amount you have paid them. You should also file a W-2 for your child at the end of the year.

Let me stress that this is all perfectly legal; small-business owners of every kind hire their kids, pay them fair wages, and reduce their own income, thus reducing their taxes. What you make your children do with their earnings is up to you—you're still the parent, and if you want, you can require them to save it all for college, pay for their own car insurance, buy their own clothes, give 10 percent to charity, or whatever your parenting philosophy deems is right.

Interns or apprentices

Many small farms provide education and training to aspiring farmers in exchange for their work. Most also provide a stipend rather than hourly wages, which generally works out to less than minimum wage. From the farmer's perspective, interns are a low-cost source of labor, and they bring youthful energy and idealism.

From the interns' perspective, spending a season working on a farm can be challenging. Most young people have no experience with farming or even gardening, and working on a small farm can be a life-changing experience for them. The farms with the most successful intern programs are those that also provide housing and meals for their workers on the farm, so that the job becomes like a summer camp experience, albeit an arduous one.

An important aspect of intern programs is the educational component. To be fair to these young people, farmers must set aside time not only to train them in the skills they need immediately, but also to educate them about the bigger picture in farming. They should learn a fair amount about ecology—how the natural world works and how farming intersects with natural processes. They should learn about farm finances—how much it costs to produce food, how prices are set, how the farmer is compensated. The young person should complete the farm internship with a good sense of why farming is important and some exposure to all aspects of farming life, including the pleasures of sharing good food, working together companionably, and the satisfaction of a job well done.

In several parts of the country, farmers have set up cooperative educational programs for interns. Interns occasionally go to other farms for farm tours and community suppers, giving them the chance to see how other farms work and to get together with a peer group socially.

One of the best places to find interns is through a local college, especially one with an environmental studies program. In general, agriculture programs do not produce the kind of young people who want to work on a small farm. Liberal arts students are better candidates. If you find a professor who's interested in your work, so much the better. He or she may be able to arrange for college credit for students who apprentice for you, making the job even more attractive.

Several other programs and organizations offer intern listing services: ATTRA, Willing Workers on Organic Farms (WWOOF), Tilth, and Maine Organic Farmers and Gardeners are just a few. Additionally, there is a program to bring interns from other countries to work on sustainable farms. Multinational Exchange for Sustainable Agriculture (MESA) is a nonprofit organization that sponsors annual eight- to 12-month on-the-farm training programs for global farmers, matching them with US host farms practicing organic and/or sustainable agriculture. The MESA program offers practical, vocational training that exposes young aspiring farmers to every aspect of small-farm management. US farmers provide hands-on training, housing, meals, and monthly program fees for the opportunity to host an experienced international trainee.

The very best interns, from abroad or from the United States, are young people who are considering farming as a career. For

What Interns Want

Here are some tips for having a successful internship program:

* **Make the rules clear.** Tell the interns when they are expected to start work, end work, and how long their breaks should be. Have policies on working hours, and don't vary them or make the intern feel guilty for quitting at 5:00 p.m. when you have to keep working until 8:00 p.m. This means you need to look ahead at your needs and be realistic about how much to expect from your interns. If work hours have to change in response to a heavier workload, give the interns plenty of warning.

* **Teach your interns a new task, then check back frequently** at first to make sure it's being done to your satisfaction. Quality can go downhill fast if you don't insist on perfection at the start. Flaws that may be obvious to your experienced eyes need to be pointed out to an inexperienced worker.

* **Pay enough.** Some farms listed on the ATTRA website don't pay any stipend, just room and board. Others list stipends ranging from $300 to $700 per month—well below minimum wage in most cases. People who aren't being fairly compensated may not work as hard or feel as committed as they would with a decent wage.

* **Some farmers house interns in their own home**, which can create a lot of stress for both parties. Separate living quarters are preferable, if possible. Rules about when the interns can be in the house for cooking and laundry need to be clear. Rules about their own living quarters should also be stated in advance, particularly with regard to drugs, alcohol, and visitors.

* **Feed them well.** You can require that interns take turns cooking, you can hire a cook, or you can cook for them. But remember that hard physical work requires good nutrition, and be sure your interns are able to eat well.

* **Tell interns what to bring with them:** for example, rain gear, boots, work gloves, water bottle for the field; blankets, towels, sheets, alarm clock for their sleeping quarters.

* **Establish safety standards for interns.** Remember that most of them will have no experience with country living and will need some guidance to stay healthy. Some farms require interns to wear shirts and sunscreen, a lesson learned from losing too many work hours to sunburn. Require shoes when they're working with tools. (You wouldn't believe how many interns think they can go barefoot.) Teach them to recognize the signs of heat exhaustion, how to check for ticks, where to watch out for snakes, and so on.

* **Have a weekly meeting** to hash out problems and answer questions.

* **Expect them to be sore at first.** Farming uses muscles most people don't normally use. Interns may need instruction in how to hold a hoe or use a shovel. Be patient; give them time to get in shape.

* **Don't be stingy with praise.** Compliment them on a job well done, let them know how important they are to your farm, and they will continue to work hard for you.

them, working on a farm is serious business, and they are eager to learn as much as possible. These folks are few and far between, however, and the majority of intern candidates will want to work on a farm for some combination of curiosity and idealism. In either case, interns don't usually have much experience when they start work, and they usually don't stay for more than a few months.

Some farms use an end-of-season bonus to encourage interns to stay longer—only those who are still working on the farmer's target date will get the extra pay.

One caution about hiring interns/apprentices: This is a murky area legally. The federal Fair Labor Standards Act provides for the employment of certain individuals at wage rates below the statutory minimum. Such individuals include student-learners (vocational education students). Such employment is permitted only under certificates issued by the US Department of Labor's (DOL) Wage and Hour Division. However, I don't know any farms that have gone to the DOL for a certificate. I do know of one farm that was visited by Maryland officials and then declared a migrant worker camp because student interns were living in cabins on the farm. As a migrant worker camp, specific housing standards apply, such as the size of the sleeping quarters, number of bathrooms, and so on. This farm did not meet those standards and was subsequently forced to fire their interns; they went out of business as a result.

Most farms are flying under the radar with their interns and apprentices. They don't ask too many questions about whether labor laws apply, probably because they don't want to know the answers. According to Neil Hamilton, a law professor at the Drake University Agricultural Law Center, it doesn't really matter whether farmer and intern enter into an agreement regarding hours, wages, and so on. This would not necessarily protect the employer later if it were determined the intern was an employee. In general, the more the relationship looks like a form of employment, the more likely it is that the minimum wage law and other labor laws apply. If the job includes education and training, the more likely it is the relationship will be considered something other than employment.

Although you may think you are providing interns with a fair deal by giving them an educational experience plus a stipend, a place to stay, and all the vegetables they can eat, you can get in trouble for doing it. I know a couple of farmers who were shocked when interns filed complaints with state authorities. In one case, interns complained that they should have received minimum wage—even though they had signed an agreement with the farmer that they were accepting lodging and food as part of their compensation. The farmer said she learned her state would be satisfied if she paid minimum wage but withheld an amount for room and board. The net pay was the same, but it complied with the letter of the minimum wage law.

An excellent resource is the 2013 publication *Farmers' Guide to Farm Internships* from the Farmers' Legal Action Group, www.flaginc.org.

H-2A program

The US government's H-2A program has made it possible for farmers to bring people from other countries to work on farms when there is a labor shortage. Many small farmers have established long-term relationships with residents of other countries who come to the United States seasonally. Although the program is criticized for its red tape and obscure rules, it has nevertheless provided a vehicle for small farms to grow their business with the help of capable workers.

The H-2A program requires employers to prove that there are not enough US workers available to fill the job openings; the employer must advertise the job and list it with the state employment service. The employer must pay transportation for the workers to and from their country. And the workers must be paid the prevailing wage, the same wage that would be paid to US workers.

However, with Congress working on major revisions to immigration law, the future of the H-2A program is uncertain. Some of the proposals being considered would impose harsh penalties on employers who hire illegal immigrants, even unknowingly. This is an area where farmers—both new and established—should tread carefully.

Local workers

In many states, agricultural employers are subject to different employment laws than other employers; in general, the laws are less restrictive for farm work. However, labor law varies from state to state on many matters, including whether you have to pay unemployment compensation, workers' compensation, the minimum wage, and so on. You will also find differences among the states in laws pertaining to hiring children other than your own. Call your state labor department or Small Business Development Center to be sure you're familiar with the rules.

Aside from all the federal and state regulations, the million-dollar question for most farmers is where to find good workers. Some say they have great luck advertising in the local newspaper, even though it brings a flood of inquiries. Others say newspaper ads bring the wrong kind of candidates and are a waste of time. They prefer to target certain audiences by placing ads in such places as colleges, natural foods stores, and environmental group newsletters. You'll just have to experiment to find out what works for you. Be sure to check references. The traits that will help you, even more than experience (which few people have today, anyway), include work ethic; physical stamina; eagerness to learn something new; a positive, can-do attitude; and the ability to be outdoors all day.

Deciding on Insurance

One of the business matters that farmers must think about, like it or not, is insurance. There's not much enjoyment to be had in visualizing the possible disasters that could befall the farm, and nobody gets excited about spending money on insurance, but few would deny that insurance is a necessary evil for direct-market farms.

But working your way through the insurance maze is not easy. You need insurance, but what kind? How much? And how much should it cost?

There's no simple answer that will work for every farm, because every farm is different in the amount of risk it faces and the amount of assets it has to protect. But it helps to know what's available before deciding whether you've got the right stuff, and enough of it. If your business has grown or your marketing has changed since you bought your insurance, now might be a good time to reconsider whether your insurance has you covered.

There are four types of insurance that farmers need to think about (besides the personal issues of health, disability, and life insurance): farm owners' insurance, which includes property and liability; product liability; employee coverage; and vehicle insurance. Your best resource for figuring all this out is your insurance agent. Although he or she is in the business of selling insurance, if you've got a good one, you can trust the advice you receive.

The most important thing to know about talking to an insurance agent is that you have to be completely honest about every aspect of your business and make the agent understand exactly what it is you do. Neil Hamilton, director of the Agricultural Law Center at Drake University, advises growers not to understate any aspect of your operation in the hope of saving money on the premium.

"If you don't disclose the full nature of your business, there is a greater likelihood that the insurance you buy will be inadequate," Hamilton says. "Then, if something happens and you ask the insurer to cover you (which is why you bought insurance in the first place), you may find out your policy does not cover the situation. Then you are in the worst possible situation—you have paid good money for an insurance policy which was not what you needed, and now you have a problem for which you are uninsured."

Before you go to see an agent and explain your business, though, it helps to know some of the basics about direct-marketing risks and policies.

Some farmers don't buy insurance because they don't expect to get sued. Their operations may be small, they may not have people out to the farm, or they may feel they know their customers and don't worry about them suing. That's the optimistic view, and there are two things you need to know before you decide to adopt it.

The most important thing to consider is that someone who is injured on your farm or by your products may be forced to sue you by his or her own insurance company. They may like you, even love you, but they have signed an insurance contract that allows their company to seek repayment from you if they get injured on your farm. This is known as "subrogation." Neil Hamilton describes how it works in his book *The Legal Guide for Direct Farm Marketing*:

> Consider a situation . . . where a CSA member was injured on the farm. If Jimmy breaks his leg on the CSA his family will go to their insurer who will pay the medical expenses based on the insurance contract for first person coverage. But the insurer will also ask, "How did Jimmy break his leg and where did it happen?" Under the "subrogation" clause in the policy, the company has a right to seek recovery from someone else if they are responsible for what happened to Jimmy. If the company believes such recovery is possible, they could sue the owner of the CSA to recover from the owner's insurance (or sue the owner personally if there is

no insurance). Under the subrogation clause the company can ask their insured to be a "use plaintiff" so the suit will be in their name. Insurance companies usually don't bring suits in their own names because it might prejudice the jury. The insured is obligated to cooperate with the subrogation and to help with the case, such as by testifying. If the insured party refuses to sign or cooperate, because the third party being sued is a friend, the company can refuse to pay the coverage or seek repayment from the insured. It is important to understand for this reason you cannot depend on the fact you deal with your friends, to assume they won't sue you if something goes wrong. In most cases they will not be making this decision, the insurance company will, and the insurance company is not interested in friendships.

That's probably enough to scare you into calling an insurance agent, but if you're a gambler, you might also want to know the frequency of lawsuits against direct marketers. Charlie Touchette, director of the North American Farmers Direct Marketing Association (NAFDMA), studied direct-marketing insurance for several years while creating a policy specifically for direct marketers (more on that later), and he says there's just no industry data on direct-marketing claims. "The whole thing of direct-marketing is relatively new—even 20 years is new when you're talking about insurance statistics," he said.

Touchette looked at the claims history in the cut-your-own Christmas tree business—surely one of the most potentially dangerous direct-marketing ventures—and found "they were far less significant and fewer of them than the insurance companies imagined," he said. The NAFDMA also surveyed its members for anecdotal information about insurance claims and found only a handful. The biggest was "a badly twisted ankle at a pick-your-own apple orchard that turned into a $40,000 settlement," Touchette said.

He has also managed the liability insurance coverage for the farmers markets in Massachusetts, and in 12 years, with 50 to 70 markets covered each year, there have been only three successful claims. All three involved wind accidents—signs or canopies blowing over and hitting customers; all three customers went to the hospital; and the claims were settled for $12,000, $26,000, and $32,000.

In 1999 the NAFDMA created its own insurance program for direct-marketers, and in the next two years, claims totaling a half-million dollars were filed against 13 farmers. All the claims were on farms, not at farmers markets, and most were on entertainment farms. Some examples of the accidents that resulted in claims:

* At a Pennsylvania farm with a corn maze, business was so good that a field of corn stubble was used for overflow parking. The field caught fire, and 31 cars were destroyed.
* A customer was injured at a "haunted farm" in Montana when she fell off a fake bridge when a monster chased her.
* A child fell off a hay-wagon ride and was run over. The child was life-flighted to a hospital and, fortunately, recovered.

The number of claims was not considered unusual, but the losses exceeded revenue, and the NAFDMA insurance program was discontinued.

Although the statistical risk of an accident is probably small, accidents do happen and farmers and markets do get sued. For about a $300 annual premium, farmers don't have to worry about accidents or defending themselves against a lawsuit, but that's a personal decision.

► FARM LIABILITY POLICY ◄

If you've decided you'd better have insurance, the first policy to consider is your liability policy. Many growers, when they first sell produce, assume that their homeowner's insurance will cover them both on the farm and at a farmers market. That may or may not be true. Your homeowner's policy will cover an accident on the farm to a guest or visitor, but once that guest is paying you for your products, the relationship changes. For example, if you let a friend pick a bouquet on your farm, injuries would be covered by your homeowner's policy. If you charge that friend $20 to pick flowers, it might not be covered. Some companies won't quibble about small commercial transactions, but if you're making more than a few hundred dollars in farm sales, you'd better check to find out whether that business is covered. In some cases, you can just add excess liability coverage for your business activities. If you're currently buying only a homeowner's policy, read it carefully for mention of commercial activity, particularly the exclusions, and have a talk with your agent.

Once you start farming in earnest, you need a farm liability policy, which will cover all activities related to farming in addition

Farmers Market Insurance

Finding insurance has been difficult for many growers, often because insurance companies just don't understand the nature of farmers markets. A new insurance product is available nationwide to provide both general liability and product liability coverage for farmers market vendors. The National Farmers Market Vendor Liability Insurance Program was developed by Campbell Risk Management, a company in Indianapolis, in collaboration with the Farmers Market Coalition and several state farmers market associations.

The policy provides coverage of $1 million per occurrence with no deductible, with a $2 million annual aggregate limit. It costs $250 to $425 per year, depending on the state of residence and gross estimated annual sales, according to Larry Spilker of Campbell Risk Management. A vendor can sell at an unlimited number of farmers markets under the policy and can also cover sales to supermarkets, restaurants, schools, and other markets. The policy does not replace existing homeowner's or farm policies.

Spilker said that it took a year to research the issue of farmers market insurance and to persuade an insurance company to underwrite the policies. What he found was that although farmers market claims are not common, they do happen. General liability claims, such as when a customer trips and falls or gets hit with a windborne canopy, are more common at farmers markets. Product liability claims, such as a customer getting sick from meat or produce, also have occurred and can be devastating for a small producer.

Campbell convinced the underwriting companies that the "premium pool"—the number of vendors who will buy the policy—would be sufficient to cover claims. The demand for this kind of insurance is likely to increase in coming years because of a trend by markets to require their vendors to be insured. As an example, Spilker said of a farmers market in Houston, "Their insurance company issued an ultimatum that said if they did not require vendors to carry their own insurance and name the market as additional insured, they were going to cancel the policy for the market."

The program, launched in 2009, has been purchased by thousands of farmers, he said. Campbell Risk Management has also developed an insurance policy that covers the market itself and a policy that protects directors, officers, and staff of farmers markets. You can learn more about both insurance products as well as general issues pertaining to farmers market insurance at www.farmersmarketcoalition.org. You also can phone Larry Spilker of Campbell Risk Management at 800-730-7475, ext 203; or email lspilker@campbellrisk.com.

to the usual liabilities of owning property. Whether your direct-marketing activities are included in the company's definition of farming activities will vary, particularly if you're buying from a company that does most of its business with traditional farmers.

Again, read the exclusions to find out if roadside markets, off-farm farmers markets, and pick-your-own (PYO) operations are covered. Generally, PYOs will require additional coverage because the exposure, or potential for someone to be injured, is greater when there are more people visiting the farm.

Farm liability policies may contain two types of coverage: personal liability and medical payments to others. At my farm, for example, my insurance company would pay up to $1,000 to any person who was injured on the farm if we had not been negligent. If the injured person decided to sue us, alleging negligence, we would be covered up to $500,000, and the insurance company would handle the defense.

➤ HOW MUCH COVERAGE? ◄

This brings up the point of how much coverage you need to purchase. The old insurance maxim is "cover your assets." In theory, if someone is injured seriously because of your negligence (in the eyes of the court), the damage award could take everything you own and even attach your future earnings. In the Northeast and on the West Coast, where the price of real estate is high, and on farms with a lot of buildings and equipment, many direct-marketers insure for $2 million or more. Farmers of more modest means might decide to go with $1 million or $500,000 coverage.

The difference between premiums for $500,000 and $1 million coverage is relatively small. It would cost more to increase the medical payment for nonnegligence accidents from $1,000 to $5,000 than it would to increase liability coverage to $1 million because the risk of a small injury is greater than the risk of a big, lawsuit-producing one.

My insurance agent tells me it's unlikely that a court would force a farmer to sell the farm, but cash assets would be an easy target for the opposing attorney. And there have been cases where defendants' homes, while not taken away from them, have been put in a trust that reverts to the injured person upon the death of the owner.

➤ COVERING EMPLOYEES ◄

Your employees should be covered for injuries one of two ways: either through your state workers' compensation program or through your liability insurance.

As mentioned in the previous section, state laws vary on whether you must purchase workers' comp insurance, so you should call your state labor department to find out. In some states, you aren't required to buy it, but you can opt to. In either case, you should view workers' compensation insurance as a benefit both to your employees and business. Obviously, the employee benefits because in the case of an injury there is a standard recourse to compensation. You, as the employer, benefit as well, because if the farm does have workers' comp, an injured employee is limited to workers' comp as the sole source of recovery. That means the employee can't sue you seeking greater damages or huge punitive damage claims.

However, workers' comp insurance is expensive compared to the cost of adding employees to your farm liability policy. Our insurance agent estimated that we would pay $400 to $500 a year to participate in the workers' compensation program in Kansas. Adding four employees to our liability policy costs an extra $80 a year.

The negative side to covering employees through the liability policy is that if the employee is injured, he or she would have to sue, and your insurance company might decide to defend the suit. You would, under your contract, be obliged to help with the defense. If you care about your employees, you would want to see them compensated for injuries, not forced to face you in court to fight about it.

❧ PRODUCTS LIABILITY ❧

Your general farm liability policy may or may not cover an incident in which your farm products make someone sick. Check to find out. If you're selling fresh produce only, you're probably covered. If you're doing any value-added products, you may need to purchase separate products liability coverage.

Some stores won't buy from you unless you have a products liability policy. Some insurance companies won't even insure for farm-made products like jams, salad dressings, baked goods, and so forth, so you may need to shop around to find coverage.

❧ VEHICLES ❧

It goes without saying that you need to tell your car insurance company if you have employees driving your vehicles. In any vehicle accident, the vehicle insurance is considered the primary insurance policy that handles claims first. If someone other than the people

Crop Insurance

Until fairly recently, market farmers could not purchase crop insurance the way traditional row-crop farmers do. Federal crop insurance programs were limited to such commodities as corn and wheat. Then a program was developed in Pennsylvania to insure direct-market, specialty-crop farmers, and this has since expanded to 35 states. Called AGR-Lite, the insurance program protects a farm's total revenue rather than specific crops. The amount of income that can be covered is based on a farm's previous five years' worth of income as reported on the IRS Schedule F. AGR-Lite provides protection against low revenue due to unavoidable natural disasters and market fluctuations. Several levels of coverage are available. Producers can choose to insure 65, 75, or 80 percent of their annual income. If their income drops below that chosen percentage, they can choose to be compensated at a rate of either 75 percent or 90 percent. Premiums range from about 1 to 3 percent of gross revenue.

Here's an example of how AGR-Lite works. Farm A has an average annual gross revenue of $100,000. The farmer signs up in March for an AGR-Lite insurance policy at the 80 percent coverage level and 75 percent payment rate, and pays a premium of about $2,500. A severe drought that summer reduces gross income by half, to $50,000. Because the farmer chose the 80 percent coverage level (the highest available), $80,000 of revenue is covered. Actual revenue of $50,000 is $30,000 short of the covered level. The insurance policy will pay 75 percent of the shortfall, or $22,500. The farmer still has a bad year, but the blow is lessened by the insurance payment.

named on your policy is driving at the time of the accident, you could be in trouble.

⏵ WHERE TO FIND INSURANCE ⏴

In the past few years, farmers have been complaining about being dropped by insurance companies and being unable to find others that will cover their business activities. Again, your best bet is to sit down with an independent insurance agent and explain your business thoroughly. Coleen Perry, an insurance broker at InterWest Insurance Services in Sacramento, California, says the problem with finding insurance may be a computer glitch—she discovered that insurance companies list outdoor markets in the same category as amusement parks. Once the insurance agent clears up that misconception, several national insurance companies will gladly cover farmers markets and farms.

Many small farmers use Farm Bureau Insurance. By working as a group, the Missouri Farmers Market Association was able to secure liability insurance for its farmers markets for only $25 per market per year. The policy provides $1 million liability coverage for each market. Working together as an association was the key to striking the deal. Farm Bureau bills the association, which then bills its members. The Missouri official who negotiated the deal advises groups in other states to contact the state office of Farm Bureau about setting up a similar program.

Where to Learn More

At the beginning of this book, I said that market gardening is one of the most complicated and challenging jobs you can hold. I hope that the wide array of topics covered in the previous chapters has given you a good sense of how true this statement is. In a single day, a market gardener might do all these things: train an employee; talk to a chef; start seeds; identify pests and figure out what to do about them; harvest, wash, and pack vegetables; drive a tractor; fix equipment; calculate payroll; make signs for farmers markets; and order supplies. The market gardener exercises nearly every part of the body and brain on a daily basis.

As you've read the previous chapters, you have probably encountered some topics you already know a lot about and some that you have no experience with. You may be good with equipment, for example, but have no idea how to make connections with a chef. You might be great at marketing but not know how to choose varieties and schedule production. And so on. Everyone brings some strengths and some weaknesses to the work of market farming. My objective in writing this book is to identify the major skills required of a market farmer and to tell you where to go for more help on the topics where you need additional information.

In this final part of the book, I'll mention some of the best overall resources for beginning growers, including the books I consider essential. I'll also provide the contact information for specific resources mentioned in the previous chapters. I have three reasons for listing all the contact information here. First, addresses and websites change frequently, and compiling them into one section will make future revisions easier. The second reason is that web addresses really mess up the layout in a book. The third reason is that I plan to put this entire section on my website so that you will have one handy access point with live links to all the sites mentioned. Go to www.growingformarket.com and click the button for Market Farming Success.

Growing for Market

In the 20-plus years that I have been publishing *Growing for Market* (*GFM*), I have been told over and over how helpful it has been to beginning farmers. There is really nothing else like it because every article is written by a market farmer and provides detail that only those with experience would know. We currently publish 10 issues per year, with an average of six articles per issue. Most years, we have more than 30 writers. When I get a proposal for an article, my first consideration is, "Can readers use this information now?" If the answer is no, if there is perhaps some detail missing like pricing or costs, the article goes back for revision. It is this attention to practical details that has made *GFM* the most respected magazine for market gardeners.

Unfortunately, we can't keep everything in print forever. We try to keep at least the previous five years' worth of issues, or collected articles, on hand for those new to *GFM*. We offer a reduced price for these collections, which include a current subscription. We call it the Beginner's Special, and it continues to be a valued resource for new growers. We also offer a membership to our online archive of articles, which includes everything published from 2001 to the present. That's called a Full Access membership.

Growing for Market also sells a select number of market gardening books. We sell only the books that we consider relevant and useful, and we keep them in stock and ready for immediate shipment. Our books cover every facet of market gardening, including vegetables, flowers, herbs, fruits, marketing, farmers markets, greenhouses, hoophouses, and more. Book sales are an important component of *Growing for Market*'s financial viability, so we really appreciate the readers who buy from us.

To learn more about *Growing for Market* or to order subscriptions or books, please visit www.growingformarket.com, or phone toll-free at 800-307-8949.

Books about market farming

For those just getting started in market gardening, here are the books that I consider essential reading. I have listed them in order of importance.

The New Organic Grower by Eliot Coleman. This guide to small-scale, intensive vegetable production is the foundation of today's local food movement. Virtually every market farmer has read this book, so it's part of our common language. Its most recent revision was in 1995, but it is just as relevant and important today as then.

Sustainable Vegetable Production from Start-Up to Market by Vern Grubinger. This is an excellent overview of vegetable farming, with an emphasis on larger-scale production than *The New Organic Grower.* It covers all aspects of market farming, from planning to marketing, and does a great job of helping the small grower understand what is required to expand. The sections on equipment are particularly useful to new growers.

The Flower Farmer: An Organic Grower's Guide to Raising and Selling Cut Flowers. I'm the author of this book and I've been told by countless people that it helped them get started with this most profitable crop. Adding cut flowers to the crop mix doesn't require a lot of extra production expertise or costs, but it brings in a lot more revenue.

The Hoophouse Handbook: Growing Produce and Flowers in Hoop-houses and High Tunnels is a collection of *Growing for Market* articles about all aspects of high tunnels, including how to buy and build one (with photos and illustrations), how to grow various crops, and which crops are the most profitable.

The Winter Harvest Handbook by Eliot Coleman. This book is about Coleman's system for growing year-round in unheated or minimally heated greenhouses. It is detailed about materials, construction techniques, varieties, planting dates, and so on and is a great resource on season extension. A companion DVD, *Year-Round Vegetable Production*, is a video of a workshop taught by Coleman.

Walking to Spring by Paul and Alison Wiediger, Kentucky growers, describes how they use high tunnels year-round for a succession of profitable crops.

The Essential Urban Farmer by Novella Carpenter and Willow Rosenthal shares the experiences of two successful urban growers. It covers all the issues that are particular to urban farming, such as soil contamination, limited space, zoning and neighbors, security, and much more.

Food and the City by Jennifer Cockrall-King explores the growing urban agriculture movement.

Small Farm Equipment by Jon Magee. This little book is a real help for those with no mechanical knowledge. It explains the basics of how farm equipment works, how to use it safely, and how to maintain it.

ATTRA

ATTRA is the National Sustainable Agriculture Information Service. (It was originally called Applied Technology Transfer for Rural Areas and became so well known for its sustainable agriculture information that it has hung onto the acronym.) It is managed by the National Center for Appropriate Technology (NCAT), a private nonprofit organization founded in 1976, and is funded under a grant from the United States Department of Agriculture (USDA). That's a long way of saying that it's a federally funded information service for sustainable farmers.

ATTRA has compiled a huge number of extensive, free publications about topics of interest to market farmers. They are all written by staff members who have expertise in agriculture, and they usually include excerpts from magazines, newspapers, and Extension publications. Under the heading of Horticultural Crops, for example, there are 78 separate publications. These resources offer detailed information on production of specific horticultural crops, focusing on sustainable and organic production methods for traditional produce, and also introducing a range of alternative crops and enterprises. In these publications you can find information on strategies for more sustainable greenhouse and field production of everything from lettuce to trees.

ATTRA publications are free at www.attra.org. If you don't have Internet access, you can also call toll-free to 800-346-9140 (English) or 800-411-3222 (Español) to request printed copies of publications for a small fee.

ATTRA also offers the only nationwide internship listing service. If you are a farmer who wants to take on interns, you can list your farm for free with ATTRA. Use the same contact information as above.

SARE Learning Center

The USDA's Sustainable Agriculture Research and Education (SARE) program has funded the publication of many fine books over the years and now offers some of them as free downloadable PDFs. For example, *Building a Sustainable Business* is $17 in print but free as a download. SARE also publishes bulletins, grant project reports, and much other useful information. www.sare.org/publications/.

➤ CHAPTER 1 ◀
Getting Started

Conferences

I encourage you to go to conferences when getting started in market farming. You will learn from the speakers, certainly, but you'll get as much value from chatting in the hallway with other growers and visiting vendors at the trade show. Most market farming conferences are held in the winter months, and there are large ones in every region of the country. Start with the conference in your own region, and try to attend one of the big conferences farther afield. Some of these are epic events. Below is information on some of the largest conferences; others are listed on the ATTRA website, attra .ncat.org/calendar.

Association of Specialty Cut Flower Growers (ASCFG) holds a national conference every other year and regional meetings every year. www.ascfg.org.

EcoFarm Conference is the granddaddy of them all, held for more than 33 years at the beautiful Asilomar conference center on the Northern California coast. Beautiful surroundings, great organic meals, pre- and postconference bus tours, and many speakers. www.eco-farm.org/.

Great Plains Growers Conference is held in St. Joseph, Missouri, in January and features one day of intensive workshops plus two days of shorter presentations on many topics, including vegetables, fruits, cut flowers, poultry, marketing, CSA, and high tunnels, geared toward Midwestern commercial growers. www.greatplainsgrowers.org/.

MOSES Organic Farming Conference attracts more than 3,000 people to LaCrosse, Wisconsin, every winter for the largest organic conference and trade show in the United States. Workshops, organic food, nearly 200 vendors, music and dancing in the evening—it's the social event of the season for Upper Midwest farmers. www.mosesorganic.org/conference.html.

New England Vegetable and Fruit Conference is held every second year in December. www.newenglandvfc.org/index.html.

NOFA Summer Conference is held in August by the Northeast Organic Farming Association, and there are several state NOFA conferences held in winter. www.nofa.org/.

Ohio Ecological Food and Farm Conference is held in February
and features two days of workshops for organic farmers.
www.oeffa.org.

Pennsylvania Association for Sustainable Agriculture holds its
Farming for the Future Conference in February and attracts
more than 2,000 farmers, processors, consumers, and community leaders. pasafarming.org.

Southern SAWG Conference is held in January and features
two days of intensive short courses and two days of general
conference and trade show activities. It is tailored for those
in the South producing organic and sustainable food on a
commercial scale and for those working to improve local food
systems. www.ssawg.org/.

Sustainable Agriculture Conference of the Carolina Farm
Stewardship Association is held in the fall in North or South
Carolina. Three days of workshops and great local food.
www.carolinafarmstewards.org/sac/.

Educational organizations

These groups publish information, hold field days and conferences,
and serve small farmers in many other ways. Look for groups in or
near your state.

Alabama Sustainable Agriculture Network, PO Box 2127, Mont-
gomery, AL 36102-2127; 256-743-0742; www.asanonline.org.

Appalachian Sustainable Agriculture Project, 306 West Haywood
Street, Asheville, NC 28801; 828-236-1282; www.asap
connections.org.

California Certified Organic Farmers, 2155 Delaware Avenue,
Suite 150, Santa Cruz, CA 95060; 831-423-2263; www.ccof.org.

Carolina Farm Stewardship Association (CFSA), PO Box 448, Pitts-
boro, NC 27312; 919-542-2402; www.carolinafarmstewards.org.

Cascade Harvest Coalition, 4649 Sunnyside Avenue North,
Room 123, Seattle, WA 98103; 206-632-0606; www
.cascadeharvest.org.

Community Alliance with Family Farmers, PO Box 363, Davis, CA
95617; 530-756-8518; www.caff.org.

Florida Certified Organic Growers and Consumers (FOG), PO Box
12311, Gainesville, FL 32604; 352-377-6345; www.foginfo.org.

Future Harvest–CASA, 1114 Shawan Road, Suite 1, Cockeysville,
MD 21030; 410-549-7878; www.futureharvestcasa.org.

Kansas Rural Center, PO Box 133, Whiting, KS 66552; 785-873-3431; www.kansasruralcenter.org.

Kerr Center for Sustainable Agriculture, PO Box 588, Poteau, OK 74953; 918-647-9123; www.kerrcenter.com.

Land Stewardship Project, 301 State Road #2, Montevideo, MN 56265; 320-269-2105; 180 E. Main Street, Lewiston, MN 55952; 507-523-3366; 821 E. 35th Street #200, Minneapolis, MN 55407; 612-722-6377; www.landstewardshipproject.org.

Maine Organic Farmers and Gardeners Association, PO Box 170, Unity, ME 04988; 207-568-4142; www.mofga.org.

Michael Fields Agricultural Institute, W2493 County Road ES, PO Box 990, East Troy, WI 53120; 262-642-3303; 16 N. Carroll Street, Suite 810, Madison, WI 53703; 608-256-1859; www.michaelfieldsaginst.org.

Midwest Organic and Sustainable Education Service (MOSES), PO Box 339, Spring Valley, WI 54767; 715-778-5775; www.mosesorganic.org.

Nebraska Sustainable Agriculture Society, PO Box 736, Hartington, NE 68739; www.nebsusag.org.

Northeast Organic Farming Association, Box 164, Stevenson, CT 06491; 203-888-5146; www.nofa.org.

Northern Plains Sustainable Agriculture Society, PO Box 194, 100 1st Avenue SW, LaMoure, ND 58458; 701-883-4304; www.npsas.org.

Ohio Ecological Food and Farm Association, 41 Croswell Road, Columbus, OH 43214; 614-421-2022; www.oeffa.org.

Oregon Tilth, 260 SW Madison Avenue, Suite 106, Corvallis, OR 97333; 503-378-0690; www.tilth.org.

Organic Seed Alliance, PO Box 772, Port Townsend, WA 98368; 360-385-7192; www.seedalliance.org.

Pennsylvania Association for Sustainable Agriculture, PO Box 419, Millheim, PA 16854; 814-349-9856; www.pasafarming.org.

Rural Roots, PO Box 8925, Moscow, ID 83843; 208-883-3462; www.ruralroots.org.

Samuel Roberts Noble Foundation, 2510 Sam Noble Parkway, Ardmore, OK 73401; 580-223-5810; www.noble.org.

Texas Organic Farmers and Gardeners Association, PO Box 48, Elgin, TX 78621; 512-656-2456; www.tofga.org.

Tilth Producers of Washington, Good Shepherd Center, 4649 Sunnyside Avenue N #305, Seattle, WA 98103; 206-632-7562; www.tilthproducers.org.

Virginia Association for Biological Farming, PO Box 1003, Lexington, VA 24450; 540-463-6363; www.vabf.org.

Leasing land

The Landowner's Guide to Sustainable Farm Leases from Drake University Agricultural Law Center can be downloaded at sustainablefarmlease.org/the-landowners-guide-to -sustainable-farm-leases/.

▶ CHAPTER 2 ◀
The Markets

Wherever you market your produce, you should get your farm listed on LocalHarvest.org, a website dedicated to linking producers and consumers. You can create a free listing with LocalHarvest.org if you are a direct-marketing family farm, a producers' farmers market, a business that sells products made from things grown locally by family farms, or an organization dedicated to promoting small farms and the "Buy Local" movement. www.localharvest.org.

CSA resources

At this writing, I'm aware of three software platforms to help CSA farmers manage sign-up and payment, take online orders, create picking lists, and organize deliveries. All have their own features, so explore each in turn to see which works best for your business.

Member Assembler from Small Farm Central, a business that creates websites for farmers: www.smallfarmcentral.com /memberassembler.
CSAware from LocalHarvest.org: www.csaware.com.
Farmigo, www.farmigo.com.

Publications

Local Harvest: A Multifarm CSA Handbook is available free from the SARE Learning Center: www.sare.org/Learning-Center/Books.
The Kansas Rural Center's publication *Subscribing to Change* is about the Rolling Prairie Farmers Alliance, an eight-farmer CSA in eastern Kansas that has been operating for more than 15 years. It is available free on the KRC website: www.kansasruralcenter.org.

Wholesale Success: A Farmer's Guide to Food Safety, Selling, Post-harvest Handling, and Packing Produce is an excellent manual for all producer growers. It emphasizes practices that will ensure food safety and a long shelf life—essential for wholesale growers but helpful for farmers market and CSA growers as well. It's a big, heavy, spiral-bound book and is available from www.familyfarmed.org.

The US Department of Agriculture has a huge amount of information on farmers markets, food hubs, and agricultural cooperatives. Start exploring at www.ams.usda.gov/.

❧ CHAPTER 3 ❧
The Crops

Vegetables

The best place to find information about commercial vegetable production of major crops is from the state Extension services. The best way to locate publications is to use an Internet search engine with the term "commercial vegetable production" and the name of the veggie you want to find. From the results, pick the state nearest your own, but also read a couple of others to get a broader perspective on how vegetables are grown. I particularly enjoy the publications from North Carolina and Virginia.

Microgreens

Johnny's Selected Seeds has a fact sheet on micromix production, as well as a large selection of seeds that are suitable for micro-greens. 877-564-6697; www.johnnyseeds.com.

Several publications are available to explain Good Agricultural Practices and HACCP plans as they pertain to salad mix. The FDA *Guide to Minimize Microbial Food Safety Hazards for Fresh Fruits and Vegetables* can be found at www.fda.gov /downloads/Food/GuidanceComplianceRegulatoryInformation /GuidanceDocuments/ProduceandPlanProducts /UCM169112.pdf.

Cornell University has produced *Food Safety Begins on the Farm: A Guide for Growers,* located at ecommons.cornell.edu/handle /1813/2209.The Cornell University Good Agricultural Practices Network for Education and Training: www.gaps.cornell.edu.

❧ CHAPTER 4 ❧
Equipment and Tools

To get a good sense of the kinds of equipment out there, it pays to start collecting catalogs or perusing websites. You might encounter a problem and not even know a solution exists if you don't educate yourself about farming equipment and tools. Here are some of the dealers and suppliers I consider essential.

Farm Hack is an online community of farmers working to share information and experience about scale-appropriate tools. It is hosted by the National Young Farmers Coalition, an organization for anyone in the first 10 years of a farming career and can be found at www.youngfarmers.org.

Tractors and implements

The BCS is the most commonly used type of walk-behind tractor on market farms. That's primarily because there are many kinds of implements that can be used on the BCS, ranging from the basic tiller to hay balers and mowers. You can see the range of possibilities at the Earth Tools website, www.earthtoolsbcs .com; 1525 Kays Branch Road, Owenton, KY 40359; 502-484-3988. You may also find local dealers for BCS tractors.

Ferrari Tractor specializes in scale-appropriate equipment from manufacturers around the world. Walking tractors, bed shapers, greens harvesters, walk-behind combines—if they are in use in Europe, you can get them through Ferrari Tractor. 530-846-6401; www.ferrari-tractors.com.

A great source of information about vegetable planting and harvesting equipment is Market Farm Implement, a farm-based business in Pennsylvania. The company sells both new and used equipment. It has a warehouse you can visit, or you can view equipment online at www.marketfarm.com. For more information: Market Farm Implement, 257 Fawn Hollow Road, Friedens, PA 15541; 814-443-1931.

Steel in the Field: A Farmer's Guide to Weed Management Tools is a book that describes the many kinds of cultivation equipment available. It's published by Sustainable Agriculture Network. Handbook Series No. 2, Sustainable Agriculture Publications,

University of Vermont. To order, or for more information, call
802-656-5459 or email sustainable.agriculture@uvm.edu.
Vegetable Farmers and Their Weed Control Machines is a video on
the same topic. Ordering information is the same as above.

Wheel hoes

There are at least four brands of modern wheel hoes available in the
United States:

Valley Oak Wheel Hoe is made by a small-scale tool company in
California. We have used one and found it to be satisfactory
in every way. PO Box 301, Chico, CA 95927; 530-342-6188;
www.valleyoaktool.com.
The Real Wheel Hoe is a Swiss-made tool sold by several seed and
farm suppliers in the United States. It is more expensive to
purchase and to repair, but it's heavier, which may be either an
advantage or disadvantage depending on the size of the farmer.
Check with Johnny's Selected Seeds, www.johnnyseeds.com.
The Hoss wheel hoe is available from www.hosstools.com.
The Maxadyne is available from www.weedwithspeed.com.

Tools and supplies

Several companies specialize in the kinds of tools and supplies you'll
need on a market farm. This list is of companies that sell a wide variety of tools and supplies for greenhouse and field production. See
separate entries for specialized supplies named in other chapters.

A. M. Leonard supplies the nursery and landscape industry and is
a source for landscape fabric and hand tools. 800-543-8955;
www.amleonard.com.
Barr Inc. specializes in reconditioned coolers and will ship anywhere. 888-661-0871; www.barrinc.com.
BWI is a horticultural supplier with 14 locations serving much
of the Midwest, South, and Southeast. 903-838-8561;
www.bwicompanies.com.
Fred C. Gloeckner & Co. sells seeds to the commercial greenhouse
industry but also has a supply catalog with all kinds of greenhouse tools. 800-345-3787; www.fredgloeckner.com.
G&M Ag Supply. 928-468-1380 or 800-901-0096;
www.gmagsupply.com.

Gempler's carries tools and clothing for agriculture and greenhouse. 800-382-8473; www.gemplers.com.

Griffin Greenhouse Supply serves the Northeast and mid-Atlantic states. 800-888-0054; www.griffins.com.

Growers' Supply by Farm Tek. 800-476-9715; www.GrowersSupply.com.

Harmony Farm Supply focuses on sustainable and organic growers, with a full line of tools and supplies, including cover-crop seeds and irrigation supplies. 707-823-9125; www.harmonyfarm.com.

Hummert International has locations in Missouri and Kansas, and will ship anywhere else. 800-325-3055; www.hummert.com.

Johnny's Selected Seeds sells tools and supplies as well as seeds. The broadfork is available here. Be sure to specify that you want the commercial growers' catalog. 877-564-6697; www.johnnyseeds.com.

McConkey Co. is a horticultural supplier in California, Oregon, and Washington. 800-426-8124; www.mcconkeyco.com.

Morgan County Seeds is a farm-based purveyor of seeds, tools, and vegetable farming machinery. 573-378-2655; www.morgancountyseeds.com.

Peaceful Valley Farm Supply. 888-784-1722; www.groworganic.com.

Woodcreek Farm & Supply sells all kinds of fertilizers, pest control products, etc. 276-755-4902; www.woodcreekfarm.com.

Greenhouse and hoophouse manufacturers

Agra Tech. 925-432-3399; www.agra-tech.com.

Atlas Greenhouse Systems. 800-346-9902; www.atlasgreenhouse.com.

BWI Companies. 903-838-8561; www.bwicompanies.com.

Conley's Greenhouse Manufacturing & Sales. 800-377-8441; www.conleys.com.

Farm Tek's Growers Supply. 800-476-9715; www.GrowersSupply.com.

G&M Ag Supply. 928-468-1380 or 800-901-0096; www.gmagsupply.com.

Harnois C.P. 450-756-1041; www.harnois.com.

Hummert International. 800-325-3055; www.hummert.com.

Jaderloon. 800-258-7171; www.jaderloon.com.

Keeler-Glasgow. 800-526-7327; www.keeler-glasgow.com.

Ledgewood Farm. 603-476-8829; www.ledgewoodfarm.com.

Ludy's Greenhouse Manufacturing. 800-255-5839; www.ludy.com.

McConkey. 800-426-8124; www.mcconkeyco.com.

Nexus. 800-228-9639; www.nexuscorp.com.

Oehmsen Midwest. 800-628-4699; www.oehmsen.com.

Paul Boers Total Growing Systems. 905-562-4411;
    www.paulboers.com.

Poly-Tex. 800-852-3443; www.poly-tex.com.

Stuppy Greenhouse Manufacturing. 800-733-5025; www.stuppy.com.

X.S. Smith. 800-631-2226; www.xssmith.com.

Zimmerman high tunnels are available from Morgan County
    Seeds. 573-378-2655; www.morgancountyseeds.com.

➣ CHAPTER 5 ❧
Planning Your Production

Johnny's Selected Seeds offers a plant calculator that tells you both the number of weeks to start seeds before setting them out and the same time to set out plants relative to the frost-free date: www .johnnyseeds.com/e-pdgseedstart.aspx.

Day length

As you start to develop a planting calendar, it's helpful to be aware of your day length in any given month. Here's a link to a great calculator that allows you to choose from a drop-down list of cities to find your latitude, longitude, and the current day length: www.exptech .com/sunrise.htm.

Record-keeping

AgSquared is online farm management software that helps you
    create a yearly plan and monitor your progress through the
    seasons: www.agsquared.com/.

Farmigo is a national network of food communities bringing the
    community-based farmers market online: www.farmigo.com/.

COG Pro provides record-keeping software simplifying documen-
    tation of fertilizers, weed and pest control methods, and other
    farm inputs for organic certification: cog-pro.com/index.html.

GAP Pro is a web-based record-keeping service that helps farmers
    to track their use of production and processing protocols known
    as Good Agricultural Practices (GAPs): www.gap-pro.com.

Dan Kaplan of Brookfield Farm in Massachusetts uses computer
    spreadsheets on Microsoft Excel for crop planning and

record-keeping on his CSA farm. The disks with the spreadsheet templates can be obtained by sending a donation of $25 to Brookfield Farm. Contact Dan Kaplan, Brookfield Farm, 24 Hulst Road, Amherst, MA 01002; 413-253-7991; bfcsa@aol.com; www.brookfieldfarm.org.

Crop enterprise budgets

Publications on high-tunnel tomatoes and melons are available from University of Missouri Extension for $10 each. To order: MU Publications, 2800 Maguire, Room E1, Columbia, MO 65211; 800-292-0969; extension.missouri.edu/publications/index.aspx.

➤ CHAPTER 6 ◄
Planting and Tending Your Crops

Greenhouse potting mix recipes

Ohio Ecological Food and Farming Assn. www.oeffa.org /editorscorner.php.
ATTRA. www.attra.org (see p. 236 for more information on ATTRA).
Compost-based seed-starting mix that has won fans in the Northeast is now available nationwide from Vermont Compost Company. 802-223-6049; www.vermontcompost.com.

Plug suppliers

Several seed companies serve as brokers for the numerous plug producers that are located around the country. Here are three that will send you a plant catalog:

Germania Seed Company. 800-380-4721; www.germaniaseed.com.
Gloeckner & Co. 800-345-3787; www.fredgloeckner.com.
Harris Seed Company. 800-544-7938; www.harrisseeds.com.

Deer control

A publication about deer management on the farm: www.ext .colostate.edu/pubs/natres/06520.html.
Suppliers of deer fencing include:

Benners Gardens. www.bennersgardens.com.
Deer Resistant Landscape Company. www.deerxlandscape.com.

G&M Ag Supply. 800-901-0096. gmagsupply.com/.

Harmony Farm Supply. 707-823-9125; www.harmonyfarm.com.

Irrigation

Some of the best publications for vegetable growers come from
North Carolina State University. They are clear, comprehensive and to the point. Here is a link to an excellent publication
on irrigation: www.ces.ncsu.edu/depts/hort/hil/hil-33-e.html.

Penn State has a publication on irrigation on small-scale vegetable
farms, complete with cost estimates for various types of irrigation systems: extension.psu.edu/business/ag-alternatives
/horticultural-production-options/drip-irrigation-for
-vegetable-production.

A good how-to publication on drip irrigation is from Kansas State
University: www.ksre.ksu.edu/bookstore/pubs/MF1090.pdf.

DripWorks is a full-service irrigation supplier, including drip tape,
sprinklers, filters, pond liners, etc. The company also can help
design a system. 800-522-3747; www.dripworksusa.com.

Irrigation Mart. 800-SAY-RAIN; irrigationmart.com/.

Rain-Flo Irrigation. 717-445-3000; www.rainfloirrigation.com.

➤ CHAPTER 7 ◄
From Field to Market

Packaging

An excellent discussion of produce packaging is available from North
Carolina State University at www.bae.ncsu.edu/programs
/extension/publicat/postharv/ag-414-8/. The publication describes
the various types of packaging, and it lists commonly used packages, weights, and quantities for all the major types of produce.

Formtex sells clamshells and corrugated boxes. www.formtex.com.

Glacier Valley Enterprises has a wide selection of packaging
supplies and containers for fruit and vegetable farmers. And it
offers smaller quantities than many produce package suppliers. 800-236-6670; www.glacierv.com.

Boxes, bushel baskets, clamshells, and so on can be purchased from
Monte Package Company. 800-653-2807; montepkg.com.

Hubert Company sells fixtures, displays, and supplies to supermarkets. You will find produce packaging, clamshells, and
much more here. Hubert is also a source for the natural plastic

containers made from corn, known as PLA, or polylactic acid. 866-482-4357; www.hubert.com.

Produce boxes, bags, mesh bags, bushel baskets, and more are available from Southern Container Corp. 800-261-2295; www.socontainers.com/ProducePackaging.htm.

Marketing supplies

A. Steele Co. sells supplies for farmers market vendors, including portable scales, cash registers, wireless credit card terminals, and E-Z UP tents. 800-693-3353; www.asteele.com.

Eat Local Food produces fine-art graphics of vegetables and fruits, including banners, postcards, tote bags, and other marketing materials. 734-341-7028; www.eatlocalfood.com.

Grower's Discount Labels is a farm-based business that designs and prints custom labels for farm products. 800-693-1572; www.growersdiscountlabels.com.

Produce Promotions sells banners, flags, bags, baskets, and other marketing products. 888-575-4090; www.producepromotions.com.

❧ CHAPTER 8 ☙
Managing Your Business

The IRS has remarkably clear publications on farming tax issues. They can be found at www.irs.gov, or request the "Farmer's Tax Guide" from 800-829-3676.

For information about farmland property tax assessments, the American Farmland Trust and the USDA have compiled numerous publications on the website www.farmlandinfo.org.

How to track electricity usage: The Kill A Watt power measurement tool is available from Real Goods. 888-567-6527; www.realgoods.com.

Deciding on a legal structure for your business requires the advice of an attorney or accountant. Here is a link to a publication from Kansas State University that describes the various options: www.ksre.ksu.edu/bookstore/pubs/MF2696.pdf.

Where to advertise for interns

Most regional and state organic farming associations have newsletters, some with online listings, where you can place a classified ad for apprentices.

ATTRA, the sustainable farming information source, allows
farmers to list internships on their website:
www.attrainternships.ncat.org.

Growing for Market runs classified ads for apprentices.
www.growingformarket.com.

Willing Workers on Organic Farms is an international organization
that allows farmers and interns to advertise for each other.
www.wwoof.org.

Several organic associations offer matching services for their
member farms:

Carolina Farm Stewardship Association,
www.carolinafarmstewards.org.

Growing Growers, a program for Kansas and Missouri farmers,
www.growinggrowers.org.

Hawaii Organic Farmers Association,
www.hawaiiorganicfarmers.org.

Maine Organic Farmers and Gardeners Association, www.mofga.org.

NOFA-Vermont, www.nofavt.org.

Ohio Ecological Food and Farm Association, www.oeffa.org.

Payroll services

QuickBooks Payroll, www.intuit.com.

ADP. 800-225-5237; www.adp.com.

Local accounting offices also do payroll for small businesses; be
sure to compare prices and services.

You can type "child labor law" plus your state into an Internet
search engine to find your state's labor department.

Insurance

The company that provides our farmers cooperative's products
liability coverage (as well as our personal farm policy) is called
Goodville Mutual Casualty Company, www.goodville.com.
It's based in New Holland, Pennsylvania, and covers a lot of
direct-market farmers in these nine states: Pennsylvania,
Delaware, Maryland, Virginia, Ohio, Indiana, Illinois, Kansas,
and Oklahoma. You can call the company at 717-354-4921 to
find an agent near you who sells their policies.

InterWest Insurance Services, in Sacramento, California. 800-
444-4134; www.iwins.com.

USDA Grading Standards for Fresh Vegetables and Fruits

Reprinted with permission from Karen L.B. Gast, "Containers and Packaging: Fruits and Vegetables," publication MF-979 (Manhattan: Kansas State University Agricultural Experiment Station and Cooperative Extension Service, 1991), available at http://www.agmrc.org/media /cms/CD1_C07C95889B783.pdf.

➤ Vegetables ◄

Asparagus: Asparagus is sold by weight in the standard containers listed in Table 1. Spears may be loose-packed, or bundled vertically in pyramid crates. Vertical packing keeps the spears straight. Spears are sized by diameter and must be at least $5/16$ inch in diameter to be sold. USDA size grades are:

Small = $5/16$ inch to less than $8/16$ inch diameter
Medium = $8/16$ inch to less thean $11/16$ inch diameter
Large = $11/16$ inch to less than $14/16$ inch diameter
Very Large = $14/16$ inch and up diameter

Beans, snap: Snap beans are sold by weight and bulk-packed in bushel hampers and cartons. They are sized by diameter.

Sieve size	Diameter (to but not including)
No.1	$12/64$–$14.5/64$ inch
No.2	$14.5/64$–$18.5/64$ inch
No.3	$18.5/64$–$21/64$ inch
No.4	$21/64$–$24/64$ inch
No.5	$24/64$–$27/64$ inch
No.6 and larger	$27/64$ inch and larger

U.S. No.1 grade snap beans must have a maximum sieve size of 4; U.S. No. 2 has no upper limit. Both have a minimum diameter of $^{12}\!/_{64}$ inch.

Beets, bunched or topped: Beets are sold by weight and packed in the containers given in Table 1. They are usually sold bunched with 12 beets per bundle with tops attached, or loose with tops trimmed short or removed. Short-trimmed tops cannot be more than 4 inches long; topped beets cannot be more than ½ inch in length.

Broccoli: Broccoli is usually sold in cartons holding 14 and sometimes 18 individual heads, or bunches of stems of uniform size. Cartons weigh 20 to 24 pounds.

Brussels sprouts: Brussels sprouts are packaged in 25-pound bulk-pack cartons, or in flats holding twelve 10-ounce consumer-ready cups. They should be greater than 1 inch and no more than 2 ¾ inches in diameter.

Cabbage: Cabbage is sold by weight, in bulk or 50-pound sacks or cartons. Packages are labeled with head size:

Small = less than 2 pounds
Medium = 2 to 5 pounds
Large = greater than 5 pounds

Sometimes head size is given as the number of heads in a 50-pound container.

Carrots, bunched or topped: When carrots are bunched with the tops left on, the bunches must weigh more than 1 pound and contain at least 4 carrots. They are packed 24 bunches to a crate. Topped carrots are packed in consumer-ready 1- or 2-pound poly bags that are packed in 48-pound units. Carrots also are packed loose in bulk containers.

Cauliflower: Cauliflower is usually packed in a flat or 2-layer carton of 9 to 16 trimmed and film-wrapped heads. A size designation is usually given that corresponds to the number of heads in the carton. The number 9 heads are larger than number 16s.

Corn, sweet: Sweet corn is packed with 5 dozen ears in cartons or wire-bound crates. It is also packed in bags.

Cucumbers: Cucumbers are most often packed in 1 ⅑-bushel cartons. Size is based on diameter and length. Small cucumbers have diameters between ½ and 2 inches. Large cucumbers have diameters greater than 2 ¼ inches and lengths longer than 6 inches. If cucumbers are packed in smaller cartons, they are sold by count packs.

Cucumbers, greenhouse: Greenhouse cucumbers are packed in smaller cartons than field-grown cucumbers. They have a carton weight of 12 or 16 pounds, and often are plastic-wrapped (shrink-wrapped) to prevent water loss.

Eggplant: Eggplants packed in 20- to 23-pound cartons are packed 18 to 24 per carton. Size is designated by number per container.

Garlic: Garlic is packed in bulk or in a carton containing consumer-ready packages of 2 bulbs each. Bulk-packed garlic is sized.

Greens: Greens include collards, dandelion greens, kale, mustard greens and Swiss chard. They are packed either loose or in bunches, 12 to 24 per carton.

Herbs: There are no USDA standards for most herbs, and few industry standards for packing containers. Most herbs are packed in airtight bags to prevent wilting. They are packed in bulk, or in bunches of 6, 12, or 30 per container. It is best to work closely with the buyer when packing herbs.

Lettuce—romaine, big Boston, bibb, leaf: These leafy types of lettuce are most commonly packed in cartons of 24 heads.

Melons—casaba, crenshaw, honeydew, muskmelon: Melons of uniform size are packed in various size boxes. Muskmelons are packed in containers that can range from 38- to 41-pound half-cartons to 80- to 85-pound jumbo crates. Honeydews are usually packed in 30- to 40-pound cartons. The other specialty melons are packed in 25- to 35-pound cartons.

Okra: Okra is packed in various size containers which have a standard packed weight. Okra is usually sold by weight.

Onions: Dry onions are sold by weight, but are packed in standard weight containers and packed to a uniform size. Size is determined by diameter.

Garlic size designation		Diameter in inches
#11	Super-Colossal	$2^{15}/_{16}$ and up
#10	Colossal	$2^{11}/_{16}$–$2^{15}/_{16}$
#9	Super-Jumbo	$2^{7}/_{16}$–$2^{11}/_{16}$
#8	Extra-Jumbo	$2^{3}/_{16}$–$2^{7}/_{16}$
#7	Jumbo	$1^{15}/_{16}$–$2^{3}/_{16}$
#6	Giant	$1^{13}/_{16}$–$1^{15}/_{16}$
#5	Tube	$1^{11}/_{16}$
#4	Medium Tube	$1^{9}/_{16}$–$1^{11}/_{16}$

Onion bulb size designation	Diameter in inches
Small	1 to 2¼
Repackers or Prepackers	1¾ to 3 (60% or more 2 inches)
Medium	2 to 3½
Large or Jumbo	3 or greater

Green onions are bunched and packed 24 or 48 bunches per container, depending on size. Green onions can be sized by diameter:

Small = less than ½ inch
Medium = ½ to 1 inch
Large = over 1 inch

Oriental vegetables: Leafy and head-type oriental (Asian) vegetables are often bunched and packed into standard containers.

Ornamental gourds: There are no USDA grade standards; handling will depend on the buyer. Gourds are often sold by weight and packed in bulk bins, or sold like miniature pumpkins, 40 pounds in ½- to ⅝-bushel crates.

Peas, green and snow: Peas are packed in standard size containers as outlined in Table 1. They are sold by standard weight of the filled container.

Peppers: Bell peppers are packed by size into standard containers that have a specific filled weight. Sizes are small, medium, large and extra large. Chili peppers have no official standards for size and count. Standard packing containers are covered in Table 1.

Potatoes: Potatoes are packaged by size and by count per 50 pounds.

Pumpkins: Jack o'lantern and processing pumpkins are shipped in bulk or in bulk bins. Eating pumpkins (small pie types) may be packed in crates, cabbage cartons or sacks. Standard weight for these smaller packs is 40 or 50 pounds. Miniature pumpkins are packed in ½- to ⅝-bushel crates with a standard weight of 40 pounds.

Radishes: Radishes are packed topped or bunched with tops. Bunched radishes must be uniformly sized within the bunch. Sizes are:

Small = ½ to ¾ inch diameter
Medium = ¾ to 1 inch diameter

Potato size designation	Diameter in inches
Size A	1⅞ and up
Size B	½ to 2¼
Small	1¾ to 2½
Medium	2¼ to 3¼
Large	3 to 4¼

Potato count	Approximate tuber weight (ounces)
Under 50	15
50	12–19
60	10–16
70	9–15
80	8–13
90	7–12
100	6–10
110	5–9
120	4–8
130	4–8
140	4–8
Over 140	4–8

Rhubarb grades	Diameter	Length
U.S. Fancy	> 1 inch	> 10 inches
U.S. No. 1	> ¾ inch	> 10 inches
U.S. No. 2	> ½ inch	> 10 inches

Large = over 1 to 1¼ inch diameter

Extra Large = over 1¼ inch diameter

Rhubarb: Rhubarb is often packed in cartons or lugs of 20 pounds. U.S. grade standards have guidelines on length and diameter.

Rutabaga: Rutabagas are packed in 25- or 50-pound sacks or cartons, packed topped and usually waxed. They must be greater than 1¾ inches in diameter.

Spinach: Spinach can be packaged loose in bulk, loose in consumer-ready packages, or bunched. Bunched spinach is usually packed 24 bunches to a 20- to 22-pound carton. Cartons holding 10-ounce consumer-ready plastic bags are packed 12 to a carton.

Squash:

Winter squash includes green and gold Table Queen (Acorn), Turk's Turban, Delicata, Butternut, Sweet Dumpling, Kabocha, Golden Nugget, Buttercup, Delicious, Orange Marrow, Hubbard, Banana, Sweet Meat, Mediterranean and Calabaza. Winter squash is usually packed in bulk bins or smaller 40- to 50-pound crates, and sold by weight.

Summer squash includes Zucchini, Cocozelle, Chayote, Scallopini, Yellow Crookneck, Yellow Straightneck and Sunburst. Summer squash is packed in a variety of containers with standard minimum weight requirements. It is also sized by small and medium categories.

Sweet potatoes: Sweet potatoes are packed in containers that hold 40 or 50 pounds. U.S. grade standards cover the requirements for different sizes.

Tomatoes: Cherry tomatoes are sold in flats holding twelve 1-pint boxes or baskets. They are usually picked vine-ripe.

Plum tomatoes are usually packed in quart boxes or baskets, eight to a carton. They are also picked vine-ripe.

Mature green tomatoes are sold in bulk-packed cartons, holding approximately 25 pounds. They are sorted by size. Size designation is based on the number of tomatoes (in rows and columns) in a layer on a standard two-layer tomato lug.

Sweet potato U.S. Grade	Diameter (inches)	Length (inches)	Weight (ounces)
U.S. Extra No. 1	$1\frac{3}{4}$–$3\frac{1}{4}$	3-9	< 18
U.S. No. 1	$1\frac{3}{4}$–$3\frac{1}{2}$	3-9	< 20
U.S. No. 2	< $1\frac{1}{2}$	—	< 36

Size designation of tomatoes		Inches	
Name	Size	(min.)	(max.)
Maximum			
Large	4 × 5 and up	$3^{15}/_{32}$	and up
Extra Large	5 × 5 and 5 × 6	$2^{28}/_{32}$	$3^{15}/_{32}$
Large	6 × 6	$2^{17}/_{32}$	$2^{28}/_{32}$
Medium	6 × 7	$2^{9}/_{32}$	$2^{17}/_{32}$
Small	7 × 7	$2^{7}/_{32}$	$2^{9}/_{32}$
Extra Small	7 × 8	$1^{28}/_{32}$	$2^{4}/_{32}$

Pink and vine-ripe tomatoes are usually packednby uniform size in a two-layer lug or tray pack. They have softened enough that bulk packing causes too much bruising.

Turnips: Turnips are packed bunched with tops, with tops short-trimmed, or topped. Packing containers and weight requirements differ for each type of pack. Topped turnips are bulk-packed in mesh or poly film bags or bushel baskets, or packed in consumer-ready 1-pound plastic bags, 24 bags to a carton. Turnips with tops are usually bunched and packed in wire-bound crates or bushel baskets, and have a required minimum weight of 25 pounds.

Watermelon: Watermelons are sold by weight and usually in bulk bins. Prices are quoted per hundredweight.

➤ Fruits ◀

Apples: Apples are packed by count and weight. Apples sold by weight are usually packaged in consumer-ready 3-pound poly bags, 12 bags per carton. The apples are uniformly sized.

Apples are also sold by count, which is the number of apples of a certain diameter/size that will fit into a standard bushel carton. The larger the apple, the fewer per carton, so the lower the number designation. Apples can be bulk- or volume-filled into a carton, or place-packed into tray or cell packs in a carton. Tray or cell packs reduce the amount of injury to the fruit, but cost more because the tray and cell inserts must be purchased. Following is a summary of the fruit count and size, and packing arrangement for apples.

1. Count = Number of apples per carton or box.
2. Pack = Add the two numbers to get the number of rows per tray or layer.
3. Number per rows = First number is the number of fruit in 1st, 3rd, and 5th rows in the layer/tray. Second number is the number of fruit in the 2nd, 4th, and 6th rows in the layer/tray.
4. Pieces per layer or tray = Number of fruit per layer or tray.
5. Layers = Number of layers or trays per carton or box.
6. Size = Minimum fruit diameter for given count.
7. Paper = Size of wrapping papers if fruit is to be individually wrapped.

1.	2.	3.	4.	5.	6.	7.
Count	Pack	No. row	Pieces per layer	Layers	Size (inches)	Paper size
216	3 × 3	6 × 6	36	6	2⅛	9"
198	3 × 3	6 × 5	33	6	2¼	9"
175	3 × 3	6 × 7	35	5	2⅜	9"
163	3 × 2	6 × 7	33-32	5	2½	9"
150	3 × 2	6 × 6	30	5	2⅝	10"
138	3 × 2	6 × 5	28-27	5	2¾	10"
125	3 × 2	5 × 5	25	5	2⅞	10"
113	3 × 2	5 × 4	23-22	5	3	10"
100	3 × 2	4 × 4	20	5	3⅛	11"
88	3 × 2	4 × 5	22	4	3¼	11"
80	2 × 2	5 × 5	20	4	3⅜	11"
72	2 × 2	5 × 4	18	4	3½	12"
64	2 × 2	4 × 4	16	4	3⅝	12"
56	2 × 2	3 × 4	14	4	3¾	12"
48	2 × 2	3 × 3	12	4	3⅞	12"

Apricots: Apricots are sold by count and weight. When bulk- or volume-filled into 24-pound lugs, apricots are sold by weight. The size is designated by diameter in inches, or by jumbo, large, extra large, etc. When the fruit is tray-packed, it is given a count number, and price is based on that number.

Berries: Blackberries, blueberries, raspberries and strawberries are sold by volume in half-pints, pints and quarts. They are usually packed 12 (or sometimes 24) to a single-layer crate, flat or box. Blueberries can be labeled by size. The standard used is the number of fruit per pint.

Extra Large = Fewer than 90 berries per standard pint

Large = 90–129 berries per standard pint

Medium = 130–189 berries per standard pint

Small = 190–250 berries per standard pint

Cherries: Sweet cherries are bulk- or volume-filled into lugs that hold 18 to 20 pounds. The lugs are often lined with polyethylene (plastic) bags to preserve quality. Sweet cherries can be sorted by size. Fresh sour cherries are rarely seen in retail markets, except near production areas. They are very perishable, and most go to processors close to the production areas. There are no standard packs for sour cherries.

Grapes: Grapes are typically sold by weight in 23-pound lugs. Eastern or American type grapes are often sold by volume, in cartons filled with twelve 1-quart containers packed similar to berries.

Nectarines: Nectarines are sold by count of uniformly sized fruit in a bulk- or volume-filled lug, or a two-layer tray pack. The volume-filled lug must be at least 25 pounds, and the tray-pack averages 22½ pounds. Size designations range from the larger 50 size (number per lug) to the smaller 84 size.

Peaches: Peaches are usually sold by weight and sometimes by count. Shipping containers are packed with uniformly sized fruit, usually designated by diameter in inches. They are packed bulk- or volume-filled, or in tray-packs. If fruit is ranch-packed, then tray-packing is used to protect the softer fruit from bruising.

Pears: Pears are usually sold by count in bulk- or volume-filled cartons, wrapped in bulk- or volume-filled cartons, or tray-packed in lugs. The greater the count number, the smaller the fruit size. Each carton must contain uniformly sized fruit.

Plums and fresh prunes: Plums and fresh prunes are usually sold by weight of bulk- or volume-filled half-bushel lugs, with a minimum weight of 28 pounds. Fruit size is designated as 3 × 4, 6 × 6, 5 × 5, etc. These designations originated with an old 4-basket crate pack. The numbers designate the number of rows and columns in the top layer of the baskets. A 3 × 4 lug would have larger fruit than a 6 × 6 lug.

❧ Definitions ❧

Box or carton: Usually refers to a corrugated fiberboard container. It may be a two-piece telescoping box, or a carton that closes with top flaps. The contents can be place-packed with liners and layer dividers, or bulk-filled.

Crate: Usually refers to a wooden, wire-bound container. These are usually bulk-filled to a desired weight or, in the case of sweet corn, filled with 5 dozen ears.

Flat: Usually refers to a container that is place-packed in one or two layers, depending on the crop. Flats are also used to package produce that are packed in half-pint, pint and quart consumer-ready containers.

Lug: Usually refers to a container that is place-packed in two or three layers, depending on the crop. Lugs can also be bulk-filled. They are made of wood, corrugated fiberboard, or a combination of both. Standard dimensions are 16⅛ × 13¼ inches with varying depths.

TABLE A-1: Standard Size and Net Weights of Common Containers Used for Fresh Vegetables

Vegetable	Container	Approximate net weight (lb.)
Asparagus	Pyramid crate	30-36
	Half pyramid crate or carton	15-17
	Carton holding 16 1½-lb. pkgs.	24-25
Bean, snap	Bushel crate, hamper, or basket	28-32
	Carton	20-22
Beet		
Bunched	1⅖-bushel crate, 24s	36-40
	⅘-bushel crate, 12s	15-20
Topped	Sacked, as marked	25-50
Broccoli	Carton holding 14-18 bunches	20-24
Brussels sprouts	Carton	25
	Carton holding 12 10-oz. cups	7½-8
Cabbage	Sack, crate or carton	50-55
Carrot		
Bunched	Carton holding 2 dz. bunches	23-27
Topped	48 1-lb. or 24 2-lb. bags in master container	48
	Mesh bag, loose or as marked	25-55
Cauliflower	Flat or 2-layer carton holding	
	9-16 trimmed heads	18-24
	Long Island type crate	45-55
Chinese cabbage	15½-in. wire-bound crate	50-53
	1⅑-bushel wire-bound crate	40-45
Corn, sweet	Wire-bound crate 4½-5 dz.	42-50
	Sacks	35-40
Cucumber	Bushel carton or wire-bound crate	50-55
	1⅑-bushel carton or wire-bound crate	50-55
	Los Angeles lug	28-32
Cucumber, greenhouse	Carton holding 1-layer pack	8-10
	Carton	16
Eggplant	Carton packed 18s and 24s	20-23
	Bushel carton, 1⅑-bushel carton or wire-bound crate	30-35
Garlic	Carton or crate, bulk	20
	Carton or crate, bulk	30
	Carton of 12 pkgs. of 2 bulbs ea.	10

Vegetable	Container	Approximate net weight (lb.)
Greens	Bushel basket, crate, carton	20–25
	1⅗- or 1⅖-bushel, crate or carton	30–35
Herbs, fresh	Bulk, bunched, packed 6, 12, or 30 per carton	Varies
Lettuce		
Romaine	1⅑ bushel wire-bound crate	20–25
Big Boston	Carton and eastern carton holding 24 heads	20–24
Bibb	Carton	5–8
Leaf	Carton	10–13
Melon		
Casaba	Carton, bliss style, packed 4, 5, 6 or 8	32–34
Crenshaw	Carton, bliss style, packed 4, 5, 6 or 8	30–33
Honeydew	Flat crate standard	40
Muskmelon	½-carton or crate packed 12, 15, 18, 23	35–40
	Jumbo crate packed 18 to 45	70–80
	⅔-carton packed 15, 18, 24, 30	53–55
Watermelon	Bulk bin, medium size	1,400–1,800
	Carton holding 3–5 melons	65–80
Okra	Bushel hamper or crate	30
	⅝-bushel crate	18
	Carton	18
	12-qt. basket	15–18
Onion		
Dry	Sack	50
	Sack	25
	Carton holding 15 3-lb. bags	45
	Carton holding 20 2-lb. bags	40
Green	Carton/crate holding 4 dz. bunches	15–25
	Carton/crate holding 2 dz. bunches	20
	Carton	13
Pearl	Carton holding 12 10-oz. containers	8
Oriental vegetables	Lug	25–28
	Crate	75–80
	Carton	20–22
	Wire-bound crate	45

Vegetable	Container	Approximate net weight (lb.)
Ornamental gourds	½- to ⅝-bushel crate	40
	Bulk or bulk bins	900–1,200
Pea		
Green	Bushel basket or wire-bound crate	28–32
Snow	Carton	10
Pepper		
Green	Bushel carton	25–30
	1⅑ bushel wire-bound crate	25–30
Chili	Carton	27–34
	Lugs or carton, loose pack	16–25
Potato	100-lb. sack	100
	50-lb. sack or carton	50
	20-lb. film or paper bags	20
	5 10-lb. film or paper bags	50
	10 5-lb. film or paper bags	50
Pumpkin	Bulk	Semi-load
	Bulk bins	900–1,200
	1⅑-bushel crate	40 or 50
	½- to ⅝-bushel crate	40
Radish		
Bunched	Carton holding 4 dz. bunches	25
Topped	Carton holding 24 8-oz. film bags	12
	Carton holding 30 6-oz. film bags	11–12
Rhubarb	Carton or lug	20
	Carton	5
Rutabaga	Bag or carton	25
	Sack or carton	50
Spinach	Carton or wire-bound crate holding 2 dz. bunches	20–22
	Carton holding 12 10-oz. film bags	7½–8
	Bushel basket or crate	20–25
Squash		
Winter	1⅑-bushel crate	40–50
	Bulk bin carton, collapsible and reusable	800–900
	Various bulk bins	900–2,000
Summer	⅝-bushel crate or carton	21
	½-bushel basket or carton	21
	Carton or Los Angeles lug	24–28
	¾-lug	18–22
	1⅑-bushel crate	42–45

Vegetable	Container	Approximate net weight (lb.)
Sweet potato	Carton, crate or bushel basket	50
	Carton, California	40
Tomato		
Cherry	Carton holding 12 pints	16–18
Mature green	Carton	25
Pinks and ripes	2-layer flat, carton or tray pack	20
	3-layer lug or carton	30
	Carton, loose pack	20
Turnip		
Topped	Film bag	25
	Film and mesh bag or bushel basket	50
	Carton holding 24 1-lb. film bags	24

Index

✖ ✖ ✖

Note: page numbers in *italics* refer to photographs and figures; page numbers followed by *t* refer to tables.

About the Author

✖ ✖ ✖

Tracy Rasmussen

Lynn Byczynski edits and publishes *Growing for Market,* a periodical for market gardeners and farmers that has been published since 1992. She is also the author of *The Flower Farmer: An Organic Grower's Guide to Raising and Selling Cut Flowers* (Chelsea Green, 2008) and editor of *The Hoophouse Handbook.* She and her husband own Seeds from Italy, the US mail-order distributor for Franchi Seeds, an Italian heirloom seed company. They own a small farm near Lawrence, Kansas, where they grow cut flowers and Italian vegetables.

the politics and practice of sustainable living

CHELSEA GREEN PUBLISHING

Chelsea Green Publishing sees books as tools for effecting cultural change and seeks to empower citizens to participate in reclaiming our global commons and become its impassioned stewards. If you enjoyed *Market Farming Success*, please consider these other great books related to agriculture.

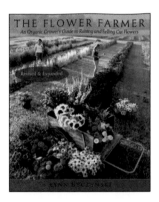

THE FLOWER FARMER, Revised and Expanded Edition
An Organic Grower's Guide to
Raising and Selling Cut Flowers
LYNN BYCZYNSKI
9781933392653
Paperback • $35.00

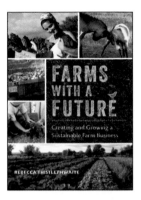

FARMS WITH A FUTURE
Creating and Growing a
Sustainable Farm Business
REBECCA THISTLETHWAITE
9781603584388
Paperback • $29.95

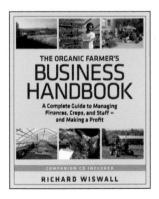

THE ORGANIC FARMER'S BUSINESS HANDBOOK
A Complete Guide to Managing
Finances, Crops, and Staff—and Making a Profit
RICHARD WISWALL
9781603581424
Paperback with CD • $34.95

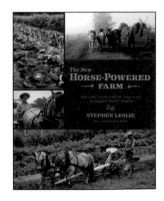

THE NEW HORSE-POWERED FARM
Tools and Systems for the Small-Scale,
Sustainable Market Grower
STEPHEN LESLIE
9781603584166
Paperback • $39.95

CHELSEA GREEN PUBLISHING
the politics and practice of sustainable living

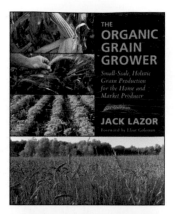

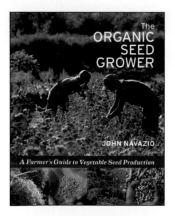
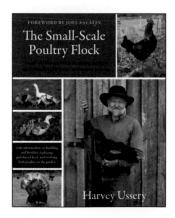